Post-Tensioning in Building Construction

Worldwide growth and application of post-tensioning in recent years is one of the major developments in building construction. The growth is propelled by a burgeoning demand for construction of serviceable and safe buildings. Unlike traditional construction, post-tensioning is based on new design methodology often not covered in traditional engineering courses.

With more than 40 years of experience of study, teaching and work on post-tensioning applications around the world, the author has written this book for students as well as practicing engineers, contractors and academics.

While the book covers the basics and concepts of post-tensioning in simple and clear language, it also focuses on the application and detailed design through real world examples.

Topics of the book include the European and the American building codes for post-tensioning design. The codes are detailed in the book's examples such as column-supported floors and beam frames. The book explains and highlights the importance of shortening specific to post-tensioned members and construction detailing for serviceable and safe performance.

Post-Tensioning in Building Construction

Bijan O. Aalami

BSc, DIC, CEng, PhD, SE(CA)
Professor Emeritus, San Francisco State University
Principal, PTstructures

CRC Press
Taylor & Francis Group
Boca Raton London New York

CRC Press is an imprint of the
Taylor & Francis Group, an **informa** business

Cover image: CCL UK

First edition published 2023
by CRC Press
6000 Broken Sound Parkway NW, Suite 300, Boca Raton, FL 33487-2742

and by CRC Press
4 Park Square, Milton Park, Abingdon, Oxon, OX14 4RN

CRC Press is an imprint of Taylor & Francis Group, LLC

© 2023 Bijan O. Aalami

Reasonable efforts have been made to publish reliable data and information, but the author and publisher cannot assume responsibility for the validity of all materials or the consequences of their use. The authors and publishers have attempted to trace the copyright holders of all material reproduced in this publication and apologize to copyright holders if permission to publish in this form has not been obtained. If any copyright material has not been acknowledged please write and let us know so we may rectify in any future reprint.

Except as permitted under U.S. Copyright Law, no part of this book may be reprinted, reproduced, transmitted, or utilized in any form by any electronic, mechanical, or other means, now known or hereafter invented, including photocopying, microfilming, and recording, or in any information storage or retrieval system, without written permission from the publishers.

For permission to photocopy or use material electronically from this work, access www.copyright.com or contact the Copyright Clearance Center, Inc. (CCC), 222 Rosewood Drive, Danvers, MA 01923, 978-750-8400. For works that are not available on CCC please contact mpkbookspermissions@tandf.co.uk

Trademark notice: Product or corporate names may be trademarks or registered trademarks and are used only for identification and explanation without intent to infringe.

Library of Congress Cataloging-in-Publication Data
Names: Aalami, B., author.
Title: Post-tensioning in building construction / Bijan O. Aalami, BSc, DIC, CEng, PhD, SE(CA), Professor Emeritus, San Francisco State University, Principal, PTstructures.
Description: First edition. | Boca Raton, FL : CRC Press, 2023. | Includes bibliographical references and index.
Identifiers: LCCN 2022051646 | ISBN 9781032307077 (hbk) | ISBN 9781032315621 (pbk) | ISBN 9781003310297 (ebk)
Subjects: LCSH: Post-tensioned prestressed concrete construction.
Classification: LCC TA683.94 .A23 2023 | DDC 693/.542--dc23/eng/20230109
LC record available at https://lccn.loc.gov/2022051646

ISBN: 978-1-032-30707-7 (hbk)
ISBN: 978-1-032-31562-1 (pbk)
ISBN: 978-1-003-31029-7 (ebk)

DOI: 10.1201/9781003310297

Typeset in Sabon
by Deanta Global Publishing Services, Chennai, India

Contents

About this book xv
About the author xvii

1 **Post-tensioning in buildings** 1
   1.1 Prestressing 1
      1.1.1 Prestressing: concept and development 1
      1.1.2 Pre-tensioning and post-tensioning 3
         1.1.2.1 Pre-tensioning 3
         1.1.2.2 Post-tensioning 4
   1.2 Growth of post-tensioning in building construction 5
      1.2.1 Early developments 5
      1.2.2 Extended development 6
         1.2.2.1 Hardware 6
   1.3 Bonded post-tensioning system 9
   1.4 Structural performance of bonded and unbonded post-tensioning systems 11
      1.4.1 Response to member deformation 11
      1.4.2 Distribution of precompression 12
      1.4.3 Response to support restraint 13
   1.5 Building code requirements 13
   1.6 Design concepts 13
      1.6.1 Load balancing 13
      1.6.2 Adoption of strip method to post-tensioning 14
      1.6.3 Adoption of single 'representative' stress for design 14
   1.7 Software application 15
      1.7.1 Strip method 15
      1.7.2 Single level finite element technology 16

    1.7.3 *Multi-level and building information modeling (BIM)-based modeling and design* 17
 Notes 17
 References 17

2 Application of post-tensioning in building construction   19

 2.1 *Introduction of post-tensioning in building construction* 19
 2.2 *Column-supported floors* 19
    2.2.1 *Flat slab construction* 19
    2.2.2 *Podium slabs* 20
 2.3 *Beam and slab construction* 21
 2.4 *Ground-supported construction* 22
    2.4.1 *Mat foundation* 22
    2.4.2 *Slabs on expansive soil* 22
    2.4.3 *Industrial floors* 23
 2.5 *Transfer floors* 24
 2.6 *Repair and retrofit* 25
 2.7 *Application in high-risk seismic zones* 26
    2.7.1 *Flat slab construction* 26
    2.7.2 *Correction of seismic deformation* 26
 2.8 *Redistribution of reactions* 28
 2.9 *Special application of post-tensioning* 33
 2.10 *Walls and frames* 33
 References 33

3 Basics of post-tensioning design   35

 3.1 *Prestressing steel* 35
 3.2 *Post-tensioning design requirements* 35
 3.3 *Post-tensioning design method options* 36
 3.4 *Straight method* 40
 3.5 *Load balancing method* 41
    3.5.1 *Simple load balancing* 44
    3.5.2 *Extended load balancing* 45
    3.5.3 *Load balancing summary* 47
 3.6 *Comprehensive method* 48
 3.7 *Application of the methods* 50
 Notes 51
 References 51

4 Building codes for post-tensioning design 53

- 4.1 Major building codes 53
- 4.2 Basic code considerations 53
  - 4.2.1 Design steps 53
  - 4.2.2 Load path 53
  - 4.2.3 Analysis schemes 54
- 4.3 Additional design requirements 54
  - 4.3.1 Deflection control 54
  - 4.3.2 Crack control 55
  - 4.3.3 Member ductility 55
- 4.4 EC2 specific code provisions 55
  - 4.4.1 Minimum reinforcement 55
  - 4.4.2 Maximum reinforcement 55
- 4.5 ACI 318 specific code provisions 55
  - 4.5.1 Minimum precompression 55
  - 4.5.2 Arrangement of prestressing tendons 56
  - 4.5.3 Transfer of column moment to slab 56
  - 4.5.4 Minimum bar length 56
- 4.6 Notable differences between EC2 and ACI 318 56
  - 4.6.1 Strength and material factors 56
  - 4.6.2 Contribution of prestressing to member strength 57
  - 4.6.3 Punching shear 57
  - 4.6.4 Application of cracking moment 58
- 4.7 European code for design of post-tensioned members 58
  - 4.7.1 EC2 code compliance basics 59
  - 4.7.2 Stress threshold 59
    - 4.7.2.1 Hypothetical extreme fiber stress 60
    - 4.7.2.2 Design crack width 63
  - 4.7.3 Load combinations 65
  - 4.7.4 Serviceability design thresholds; design limits 67
  - 4.7.5 Serviceability design flow charts 68
  - 4.7.6 Strength design 77
    - 4.7.6.1 Load combination 77
    - 4.7.6.2 Cracking moment and flexural strength 77
    - 4.7.6.3 Punching hear 78
    - 4.7.6.4 Detailing 78
- 4.8 ACI 318 provisions for post-tensioned floors 78
  - 4.8.1 Floor slab categorization and geometry 78
  - 4.8.2 Define loads 78

4.8.3	Validate sizing and material properties 78	
	4.8.3.1 Punching shear check 79	
4.8.4	Check for live load deflection 80	
4.8.5	Subdivide the slab into design strips 80	
4.8.6	Select post-tensioning and arrange tendons 81	
4.8.7	Check for average precompression 81	
4.8.8	Analyze structure; obtain design values 81	
	4.8.8.1 Analysis of structure 81	
	4.8.8.2 Extraction of design values 81	
4.8.9	Serviceability check (serviceability limit state – SLS) 81	
	4.8.9.1 Deflection control 81	
	4.8.9.2 Load combinations 83	
	4.8.9.3 Stress checks 83	
	4.8.9.4 Minimum non-prestressed reinforcement 84	
4.8.10	Safety check (ultimate limit state – ULS) 87	
	4.8.10.1 Load combinations for strength 88	
	4.8.10.2 Strength calculation 88	
4.8.11	Punching shear 89	
4.8.12	Initial condition; transfer of prestressing 89	
	4.8.12.1 Load combination 89	
	4.8.12.2 Allowable stresses 89	
4.8.13	Detailing 89	
	4.8.13.1 Tendon arrangement 89	
	4.8.13.2 Rebar arrangement 90	

Notes 90
References 91

5 Column-supported floor example   93

- 5.1 Column-supported floor 93
- 5.2 Geometry; load path; design strip 93
  - 5.2.1 Structure 93
  - 5.2.2 Design strip 93
  - 5.2.3 Design strip section properties 94
- 5.3 Material properties 96
  - 5.3.1 Concrete 96
  - 5.3.2 Prestressing 96
  - 5.3.3 Non-prestressed reinforcement 96
- 5.4 Loads 96
  - 5.4.1 Self-weight 96
  - 5.4.2 Superimposed dead load 96

       5.4.3   Dead load 96
       5.4.4   Live load 96
5.5   Design parameters 97
       5.5.1   Applicable codes 97
       5.5.2   Cover to reinforcement 97
       5.5.3   Post-tensioning system; effective stress 97
       5.5.4   Allowable design stress; crack control 97
               5.5.4.1   EC2 crack control 97
               5.5.4.2   ACI 318 crack control 98
       5.5.5   Fraction of dead load to balance; minimum precompression 98
       5.5.6   Tendon selection and layout 99
5.6   Summary of service loads 101
5.7   Analysis 101
5.8   Actions from dead loads 102
5.9   Actions from live loads 102
5.10 Actions from prestressing forces 103
5.11 Serviceability check; serviceability limit state (SLS) 105
       5.11.1   EC2 serviceability check 105
               5.11.1.1   Stress thresholds 105
               5.11.1.2   Minimum and maximum reinforcement 105
               5.11.1.3   Check hypothetical extreme fiber stresses for frequent load combination 106
               5.11.1.4   Check hypothetical extreme fiber compression stresses for quasi-permanent load combination 107
               5.11.1.5   Deflection check 108
               5.11.1.6   Crack control 108
       5.11.2   ACI serviceability check 108
               5.11.2.1   Deflection check 108
               5.11.2.2   ACI stress check 110
               5.11.2.3   ACI minimum rebar 112
5.12 Safety check; ultimate limit state (ULS) 114
       5.12.1   Strength design versus capacity check 114
       5.12.2   Calculate hyperstatic moments 114
       5.12.3   Strength check (ULS) 116
               5.12.3.1   EC2 strength check 116
               5.12.3.2   ACI strength check 117
       5.12.4   Cracking moment safety check 118
               5.12.4.1   EC2 cracking moment check 118

x Contents

     5.12.4.2 ACI cracking moment check 119
 5.13 Punching shear check 119
    5.13.1 Based on EC2 120
    5.13.2 Based on ACI[51] 123
 5.14 Initial condition; transfer of prestressing 126
    5.14.1 Load combination 126
    5.14.2 EC2 stress check 127
    5.14.3 ACI stress check 127
 5.15 Detailing 127
    5.15.1 EC2 detailing 127
    5.15.2 ACI detailing 128
 5.16 Trim bars 129
 Notes 129
 References 131

6 Design of a post-tensioned beam frame   133

 6.1 Geometry and structural system 133
    6.1.1 Effective flange width 133
    6.1.2 Section properties 136
 6.2 Material properties 136
    6.2.1 Concrete 136
    6.2.2 Non-prestressed reinforcement 137
    6.2.3 Prestressing 137
    6.2.4 Cover to reinforcement 137
 6.3 Loads 138
    6.3.1 Dead load 138
    6.3.2 Live load 138
 6.4 Design parameters 138
    6.4.1 Applicable code 138
    6.4.2 Allowable stresses 138
    6.4.3 Cracking limitation 139
    6.4.4 Allowable deflection 139
 6.5 Actions from dead and live loads 140
 6.6 Post-tensioning 141
    6.6.1 Selection of post-tensioning
        tendon force and profile 141
    6.6.2 Post-tensioning actions 142
 6.7 Code check for serviceability limit state (SLS) 143
    6.7.1 Deflection check 143
    6.7.2 Stress check/crack control 143
 6.8 Code check for strength (ULS) 147

      6.8.1   *EC2 load combination* 147
      6.8.2   *Calculation of hyperstatic actions* 147
      6.8.3   *Calculation of design moments* 148
      6.8.4   *Strength design for bending and ductility* 148
      6.8.5   *One-way shear design* 151
Notes 153
References 154

7 Member shortening; precompression: member strength   155

  7.1   *Post-tensioning; shortening; precompression* 155
  7.2   *Relationship between shortening and precompression* 155
  7.3   *Precompression and member strength* 158
      7.3.1   *No support restraint* 158
      7.3.2   *Finite support restraint* 160
      7.3.3   *Full support restraint* 161
  7.4   *Example* 169
  7.5   *Precompression in multi-story buildings* 170
      7.5.1   *Inter-story redistribution of precompression* 170
      7.5.2   *Impact of stiff walls at interior of floor slab* 171
  7.6   *Long-term shortening* 172
Notes 173
References 173

8 Stress losses in post-tensioning   175

  8.1   *Stress losses* 175
  8.2   *Immediate losses* 177
      8.2.1   *Friction* 177
      8.2.2   *Elongation* 178
      8.2.3   *Seating (draw-in) losses* 179
  8.3   *Long-term stress losses* 180
      8.3.1   *Elastic shortening of concrete (ES)* 180
      8.3.2   *Creep of concrete (CR)* 181
      8.3.3   *Shrinkage of concrete (SH)* 181
      8.3.4   *Relaxation of prestressing steel (RE)* 182
  8.4   *Examples* 183
      8.4.1   *Friction loss calculation* 183
                GIVEN 183
                REQUIRED 183
      8.4.2   *Long-term loss calculation of*
                member with unbonded tendons 185

xii   Contents

              GIVEN 185
              REQUIRED 185
    8.4.3 Long-term loss calculation of member reinforced with bonded tendons 187
              GIVEN 187
              REQUIRED 188
Note 194
References 194

9 Tendon layout and detailing   195

    9.1 *Distinguishing features in detailing PT and RC slabs 195*
          9.1.1 *Crack control and disposition of tendons 195*
          9.1.2 *Development of floor strength 196*
    9.2 *Tendon arrangement 198*
          9.2.1 *Tests on tendon arrangements 198*
          9.2.2 *Tendon arrangements in practice 199*
                9.2.2.1 Banded-distributed layout 199
                9.2.2.2 Distributed-distributed layout 202
                9.2.2.3 Banded-banded layout 204
                9.2.2.4 Irregular tendon layout 205
    9.3 *Tendon profile 205*
          9.3.1 *Common conditions 205*
          9.3.2 *Tendon at discontinuities 208*
    9.4 *Non-prestressed reinforcement 210*
          9.4.1 *Trim bars 210*
          9.4.2 *Trim bar details 214*
Other conditions 218
References 218

10 Post-tensioning construction in buildings   219

    10.1 *Post-tensioning in building construction 219*
    10.2 *System components 219*
    10.3 *Prestressing steel and strand 219*
    10.4 *Prestressing systems 221*
          10.4.1 *Bonded system 221*
          10.4.2 *Unbonded system 223*
    10.5 *Construction 224*
          10.5.1 *Quantities 224*
          10.5.2 *Construction cost 226*
          10.5.3 *Local practice 227*
          10.5.4 *Construction sequence and cycle 230*

10.6 *Stressing operation* 232
    10.6.1 *Time of stressing* 232
    10.6.2 *Stressing equipment* 234
    10.6.3 *Elongation measurement* 235
    10.6.4 *Evaluation of elongation* 236
    10.6.5 *Removal of shoring; propping* 237
10.7 *Grouting* 237
10.8 *Finishing the stressing recess* 238
10.9 *Maintenance* 239
References and Acknowledgment 240
Note 240

Index   241

About this book

This book is intended for practicing structural engineers, advanced engineering students, contractors and researchers who are engaged or interested in the design, performance and investigation of building structures using post-tensioning.

The book is the outcome of over 40 years of focused engagement in design and construction of post-tensioned structures – specifically buildings – in over 45 countries worldwide. Over the years I travelled to over 40 countries; I met the local post-tensioning engineers and contractors, learned about their practice, observed their work and exchanged views.

I learned that post-tensioning comes in many varieties, perceived and practiced in different ways. There are stout advocates of local perceptions of post-tensioning principles and their execution. Yet, as I understand it, the mechanics of post-tensioning is based on 'first principles' and is the same, regardless of local practice. This book builds on the mechanics of post-tensioning – the common thread among the diverse interpretations and practices. It introduces the basics and brings it to the point of daily application in clear and simple terms.

The book details side by side the European and the American codes for the design of post-tensioned building structures. It presents the labyrinth of the two major codes in the sequence that a design engineer needs to follow in preparing a practical design. Detailed numerical examples in the book for floor slabs and beam frames illustrate the application of the building codes covered.

The book emphasizes the importance of construction detailing in service and safety performance of post-tensioned floors, with examples and illustrations. This includes many samples of detailing and construction views.

The salient features of bonded and unbonded post-tensioning systems and the primary differences in the performance of members reinforced with each is explained in detail.

Over the years I have benefited from the valuable collaborations of Dr Florian Aalami in the design of post-tensioned structures. Specifically, our development of the leading post-tensioning design software ADAPT, now RISA-ADAPT, owes heavily to Florian's contribution. The continued effort in design and construction of post-tensioned buildings has now culminated in Visicon, which brings the design and review of post-tensioned construction to the forefront of building information modeling.

The proof reading of the book for accuracy of the numerical work relied on the talent and expertise of structural engineer Ms Narges Heidarzadh, to whom I am indebted.

Bijan Aalami, Palo Alto, February 5, 2023.

About the author

A practicing structural engineer and software developer, **Bijan Aalami**, Professor Emeritus of Civil Engineering at San Francisco State University, is an internationally renowned educator and leader in the analysis and design of concrete structures, with a specialization in post-tensioning. A native of Iran, he graduated from the University of London (Imperial College), and held positions as Professor and Vice-Chancellor of Arya-Mehr (now Sharif) University, followed by an appointment as Professor of Civil Engineering at San Francisco State University.

Bijan, a former Fulbright Scholar, has made significant contributions to understanding the behavior of post-tensioned buildings and bridges and developing procedures for their design. Bijan's decades of contribution to the post-tensioning industry were recognized through his initiation into the Hall of Fame of the Post-Tensioning Institute. A former Fulbright Scholar, he has received the American Concrete Institute's Design Award for the 'application of advanced engineering to a notable concrete structure.'

Bijan is a Chartered Engineer in the UK and a registered Structural Engineer in California. Bijan has been holding educational courses and engineering webinars throughout the United States and 45 countries worldwide.

As a Principal of the former ADAPT Corporation (now PTStructures) – a structural engineering firm in California that specializes in the design of concrete structures – Bijan has been engaged in the design and construction of concrete buildings, bridges and special structures, in particular those that are post-tensioned, for over 40 years.

Bijan was the creator and, for over 20 years, the project leader of the software suite ADAPT, now RISA-ADAPT, that is used by concrete design engineers with focus on post-tensioning in more than 75 countries worldwide.

Dr Bijan Aalami and his family live in Palo Alto, California.
bijan@PTStructures.com www.PTStructures.com

Chapter 1

Post-tensioning in buildings

1.1 PRESTRESSING

Concrete is strong in compression but weak in tension. Concrete in common construction typically resists ten times more compression than tension before it crushes in compression, or fractures in tension. Figure 1.1 illustrates the compression and tension strengths of typical concrete.

In many applications, such as in floor slabs and beams, concrete members resist the applied load primarily in bending. Under bending, the member develops compressive and tensile stresses. Since concrete capacity in tension is much less than its compression strength, the member will crack on its tension side. The cracking leads to member break-up long before the concrete draws upon its compression strength.

The common practice to overcome the shortfall in tensile strength is to use reinforcing bars (Figure 1.2). The bars are positioned where tension is likely. Tension developed in the bar through the concrete crack balances the compression in concrete on the opposing side. This leads to extended load carrying capacity of the member subsequent to its cracking.

Member cracking results in reduction of its bending stiffness. Some loss in bending stiffness is recovered through added stiffness of the reinforcement. Nonetheless, once cracked, the member will deflect more.

Addition of precompression to the member reduces the tensile stresses from bending, thereby lessening the likelihood of crack formation and larger deflections. Figure 1.3 shows the distribution of bending stress in a member and its modification through application of precompression force. Precompression reduces the tensile stresses in the member, thus enabling the member to carry larger loads before initiation of cracks.

Prestressing is the practical way for the long-term application of precompression to a concrete member.

1.1.1 Prestressing: concept and development

The concept of application of precompression to increase the load carrying capacity of a member has been known for many years. The traditional

DOI: 10.1201/9781003310297-1

2 Post-tensioning in building construction

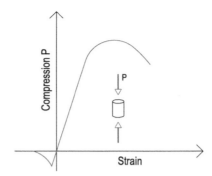

Figure 1.1 Stress-strain relationship of concrete.

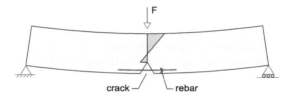

Figure 1.2 Conventionally reinforced member under reinforcement across the crack can prevent member failure.

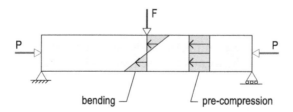

(a) Loaded member and axial load

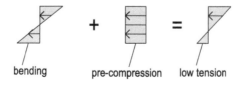

(b) Stresses in member

Figure 1.3 Member under bending and compression.

illustration used to demonstrate this concept is by holding and compressing a stack of books in equilibrium between two hands. The compression exerted by the hands prevents the books from falling. The full appreciation of its concept and its application in building construction is new, however.

1.1.2 Pre-tensioning and post-tensioning

Prestressing comes in two styles – pre-tensioning and post-tensioning. Both pre- and post-tensioning use primarily prestressing strands.

The prestressing strand is typically made of seven high-strength wires wound into a bundle. The common strands in the building construction are:

(i) 13 mm nominal diameter; guaranteed ultimate strength (f_{pk}) (1,860 MPa).
(ii) 15 mm nominal diameter; guaranteed ultimate strength (f_{pk}) (1,860 MPa).

Other sizes, both with lower and higher strength strands, are also available but less common.

1.1.2.1 Pre-tensioning

Figure 1.4 illustrates the basics of pre-tensioning steps. In pre-tensioning the prestressing strands are pulled to full force and anchored against bulkheads external to the member to be cast (part a).

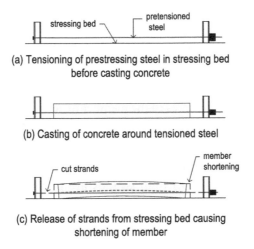

Figure 1.4 Construction stages of pre-tensioning.

Concrete is placed in the form of the member that contains the stretched strands (part b).

Once cast concrete reaches adequate strength, the strand ends extending beyond the member to the bulkheads are severed (part c). At this stage, the tension force in the prestressed steel resisted by the bulkheads is *fully* transferred to the concrete member.

Pre-tensioned members are generally manufactured off-site. Once ready, they are transported to the site and installed.

From a structural design standpoint – as expounded in the following chapters – two items in connection with the alternative scheme – namely post-tensioning – are noteworthy.

(i) At any section along the member, the compressive force in the member is equal to the tension in the prestressing steel at the same section.
(ii) The point of action of the compression force in the member at any section coincides with the center of tension of the prestressing strands.

In most applications, in anticipation of the service load, the strands are positioned off the centroidal axis of the member. The eccentricity of the prestressing force results in moment in the member. In part (c) of the figure, the strands are positioned below the centroidal axis. This provides a hugging moment that leads to member deformation shown in the figure.

1.1.2.2 Post-tensioning

Figure 1.5 illustrates the basics of post-tensioning construction. In this case, the prestressing steel is encased in a bond-breaker from the concrete surrounding it. The member is cast and allowed to gain strength before the prestressing steel is stressed against the member and anchored at the member's ends. At stressing, the member supports are detailed to allow the member to shorten under the compression applied by the jack.

A beneficial feature of post-tensioning is that the path of the tendon can be shaped to provide lateral forces to the member in addition to precompression that the member receives at its ends. The lateral forces tend the bent tendon to straighten under its pull. The straightening tendency of the tendon can be configured to counteract the anticipated loads on the member in service.

Post-tensioned applications in building construction use primarily 13 mm strands. Larger diameter (15 mm) strands are preferred in pre-tensioned members or heavier post-tensioned construction.

For example, the tendon profile shown in Figure 1.5 (c) provides an upward force along its length. The upward force opposes the gravity forces acting on the installed member.

Pre-tensioned members are generally manufactured off-site, transported and installed while the member is subject to its total prestressing force.

Post-tensioned members in building construction are commonly cast in place in their final location – often bonded to their supports and possibly

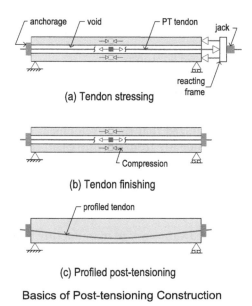

Figure 1.5 Construction stages of post-tensioning.

other members of the construction. Generally, at stressing, the post-tensioned member is a component of a larger number of members to which it is connected.

Depending on the support arrangement of a post-tensioned member, and its connection to the adjoining structural elements, the prestressing force delivered at the anchors diffuses into the member and the other parts of the structure with which the stressed member is integrated.

In the simple case shown in Figure 1.5 (c) the isolated member on rollers receives the entire compression force delivered at stressing anchors.

Depending on the support arrangement of the member and its connection to the adjoining elements, the compression in the member that houses the post-tensioning is not equal to the force in the prestressing tendons within the member. The inequality impacts the structural performance of the post-tensioned member. The impact and its mitigation are explained in Chapter 7.

1.2 GROWTH OF POST-TENSIONING IN BUILDING CONSTRUCTION

1.2.1 Early developments

Historically, in regard to the general concept and application of post-tensioning, C. W. Doehring in 1888 obtained a patent in Germany for prestressing slabs with metal wires. In 1940 Eugene Freyssinet introduced the well-known and well-accepted Freyssinet System, comprised of conical wedge anchors for 12-wire tendons. The European development was continued by Magnel in Belgium, Guyon in France, Leonhardt in Germany and Mikhilove in Russia (Nawy, E., 1997).

1.2.2 Extended development

While the contribution of post-tensioning technology in Europe is fully acknowledged, it is recognized that the development of post-tensioning in the US took place independently from the European concepts and practice. The understanding and application of post-tensioning in Europe was rooted in a different concept.

Three elements propelled the introduction and growth of post-tensioning in building construction in the US and its extension worldwide. These are:

- Simple and light hardware; easy to manufacture and handle on site by one individual.
- Simple design procedure within the skill set of common structural engineers.
- Advent and wide application of computer software.

The following is a brief highlight of each. Today's construction technology of post-tensioning is detailed in Chapter 10.

1.2.2.1 Hardware

'Deflection control' was the driving force in the introduction of post-tensioning in the US, paving the way for its extended application.

The concept was put into practice in the mid-1950s for floor slabs (Kelley, G. S., 2003). The application was to control deflection and cracking of lift-slab construction. Simply, greased strands wrapped in paper rolls, pulled and anchored at the slab edges, were used to provide precompression necessary for crack/deflection control. Figure 1.6 is the example of an early

Figure 1.6 Example of early unbonded tendon construction.

Post-tensioning in buildings 7

unbonded tendon construction of the mid-1950s in the US. Manually applied grease and paper wrapping was used for prestressing strands.

The construction did not perform well. The grease did not inhibit corrosion and the wrapping did not provide the protection needed against moisture.

The application served as an introduction to the development of the *unbonded* post-tensioning system, where the prestressing strand remains permanently detached from the concrete around it.

Over the years the system components have evolved to durable and reliable elements of building construction. In today's application the strand is coated with corrosion-inhibitive grease and encased in plastic sheathing that protects it against the elements. Billington (Billington, D. P., 2004) and Lowe (Loewe, M. S., et al.) give a perspective on the development of prestressed concrete.

Figure 1.7 shows the section of common strand and its application as a post-tensioning tendon in contemporary construction. In addition to protection against elements, the grease reduces the friction between the strand and its sheathing at stressing.

Figure 1.8 chronicles the development of unbonded tendons since its introduction in the mid-1950s.

The strands are typically secured in cast iron anchors at one end and stressed against the second anchor casting at the face of slab at the other end.

At the stressing end, the strand is passed through a conical hole in the anchor casting and seated against it using a pair of wedges. Figure 1.9 is a view of a typical anchor casting. It is used for both the stressing (live end) and the end-anchor (dead end) of the tendon.

Today's anchors, including tendon end-assembly, are well protected against intrusion of water and moisture. Chapter 10 offers additional information on hardware and construction of unbonded tendons.

Figure 1.10 shows the components of a post-tensioning system in place. These are:

(i) Prestressing tendon.
(ii) Anchor assemblies for the stressing and dead ends.
(iii) For bonded systems, vents to inject grouts in the duct and let the air out.

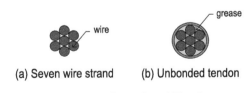

Figure 1.7 Cross-section of common strand and its application as unbonded post-tensioning tendon.

8 Post-tensioning in building construction

(a) Paper-wrapped 1955 - 1970

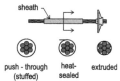

(b) Plastic sheath types 1960 - present

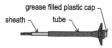

(c) Encapsulated - PTI recommended system 1985

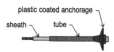

(d) Electrically isolated tendon 1983

Unbonded Tendon Evolution

Figure 1.8 Chronicle development of unbonded tendons.

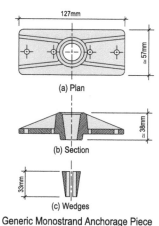

Generic Monostrand Anchorage Piece

Figure 1.9 Typical anchor piece of an unbonded tendon.

Post-tensioning in buildings 9

Figure 1.10 View of a bonded tendon slab construction ready to receive concrete.

Additional, but less critical, items are chairs or other means to set and maintain the profile of tendons during the construction and means to provide a recess at the slab edge to house the stressing anchors.

1.3 BONDED POST-TENSIONING SYSTEM

The bonded post-tensioning system for floor slabs was developed as an alternative to the unbonded system. It is used extensively worldwide – primarily outside the US.

The primary differentiating feature of the bonded system, compared to the unbonded alternative, is that in the former, the space between the strand and its sheathing is filled with cementitious grout instead of the grease used in the unbonded system.

Chapter 10 details the components of the bonded system and its construction. Briefly, the bonded tendons used in building construction are typically made up of flat ducts housing up to five strands. Figure 1.11 shows a section of a flat duct of a bonded system holding three strands. The void in the duct is filled with cementitious grout after the strands are pulled to the design force and seated.

Sheet metal or plastic ducts used for bonded tendons typically house three to five strands. The void in the duct is filled after strands are pulled and anchored.

Flat ducts are mostly used in building construction. Multi-strand flat ducts can favorably position the encased strands close to the slab surface for improved performance. This gives the strands more effectiveness to resist the applied moments. Figure 1.12 shows an example of a bonded tendon at high and low point positions in the slab. At the high and low points of the tendon in place, the centroid of the strand is offset from that of the duct by distance z.

The ducts come in flat sheet metal, corrugated sheet metal or plastic.

Figure 1.13 is a view of a floor slab reinforced with bonded tendons ready to receive concrete.

10 Post-tensioning in building construction

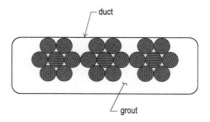

Grouted Tendon with Three Strands

Figure 1.11 Cross-section of a flat bonded tendon.

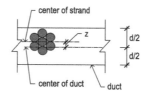

(a) Strand in duct at low point

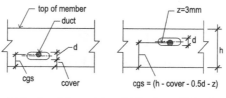

(b) Tendon at low point   (c) Tendon at high point

Center of Gravity (cgs) of Strand at Extreme Positions in Member

Figure 1.12 Examples of bonded tendon at different slab locations.

Figure 1.13 View of typical bonded tendon arrangement of reinforcement.

Details of both bonded and unbonded system hardware and construction features are given in Chapter 10.

1.4 STRUCTURAL PERFORMANCE OF BONDED AND UNBONDED POST-TENSIONING SYSTEMS

In the bonded option, the void between the prestressed steel – typically strands – and its sheathing is injected with cementitious material, namely grout. The grouting takes place after successful stressing and anchoring of the prestressing steel.

The grout bonds the encased strands and the duct housing the strands. In turn, the duct bonds with the concrete of the post-tensioned member. Consequently, through the bonds, the stretching and flexing of the strand is locked to the concrete surrounding the tendon – much like the response of non-prestressed reinforcement (rebar) under load. The deformation of the member is seamlessly transferred to that of the strand. This provides strain compatibility at the local level between the strand and its container.

1.4.1 Response to member deformation

For the bonded system, the Bernoulli assumption of 'plane sections remain plane' applies to the prestressing steel encased in concrete, as illustrated in Figure 1.14 part (a) and (c). In this case, the prestressing steel deforms compatibly with the concrete surrounding it.

In the unbonded system (parts (b) and (d) of the figure) the prestressing steel is not 'locked' to the concrete of the member surrounding it. The prestressing steel is free to slide within its duct independently from the adjacent concrete. Part (d) of the figure illustrates that the deformation of

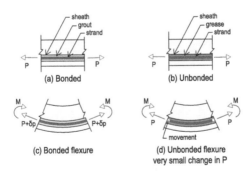

Figure 1.14 Response of bonded and unbonded member segments to flexure.

the concrete member is not seamlessly transferred to the prestressing steel. Consequently, the local flexing of the member does not translate compatibly to stress change in prestressing steel within the tendon.

1.4.2 Distribution of precompression

Prestressing strands provide precompression in the member. Precompression is beneficial to the member's crack mitigation and bending strength. This is independent from uplift that results from the tendon profile and tendon position on the section.

Figure 1.15 shows the partial plans of slabs reinforced with either bonded or unbonded tendons. Part (a) of the figure shows that at stressing, the precompression is dispersed into the slab from the tendon's anchors.

Once a bonded strand is severed, there will be local interruption in precompression at the location of the cut (Figure 1.15 (b)), but the precompression imparted from the remainder of the strand length remains essentially whole. The bond between the strand and the grout prevents the strand from fully recoiling toward the anchors and losing its extension. At a distance beyond its development length, the strand retains its tension and contributes fully to the strength and performance of the floor. The loss is local.

For bonded tendons, the question of re-stressing and re-anchoring a severed strand does not arise.

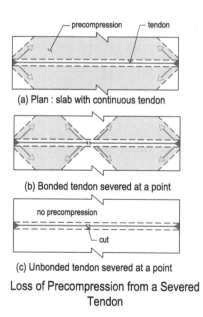

Figure 1.15 Dispersion of precompression slab severed bonded and unbonded tendon.

In contrast, when an unbonded strand is cut, the cut strand loses its force along its entire length. Thus, its contribution to both the serviceability and safety of the structure over the full length of strand is lost.

1.4.3 Response to support restraint

Post-tensioned members, floor slabs and floor beams are generally constructed on and integrated with walls and/or column supports. The supports can restrain the free shortening of the member at stressing. This is particularly pronounced at the first elevated level of multi-level constructions, where the supports may be tied to stiff foundations.

A post-tensioned member will not receive the full force of precompression from the stressed tendons, unless the member shortens under the jacking force. If restrained, a fraction of the prestressing force will be diverted to the adjoining restraining elements.

The impact of support restraint on the performance of the member is significantly influenced by whether the member is reinforced with bonded or unbonded post-tensioning, or simply conventionally reinforced. A full account of the phenomenon and the differentiating response between bonded and unbonded tendons is given in Chapter 7. Briefly, members reinforced with bonded tendons tend to be less sensitive to loss of moment capacity resulting from support restraint.

1.5 BUILDING CODE REQUIREMENTS

Both EC2 (2004) and ACI 318 (2019) cover the design of post-tensioned members, albeit to different extents. Both codes target safe and serviceable designs, but differ in the detail. The respective requirements for design of post-tensioned members and their differences are outlined in Chapter 4.

1.6 DESIGN CONCEPTS

After the introduction of simple and light hardware, the next factor propelling the wide acceptance of post-tensioning was the adoption of an easy design procedure for column-supported floors. The following explains.

1.6.1 Load balancing

The load balancing concept for post-tensioned members greatly simplified the understanding and design of post-tensioning (Lin, T. Y., 1963). The concept enabled the practicing consulting engineers to design post-tensioned floors essentially with the same skill set that they used for conventionally reinforced floors. Chapter 3 details the concept and its extension to current practice.

Briefly, load balancing provides a path to consider the post-tensioning as an applied load somewhat similar to other loads.

1.6.2 Adoption of strip method to post-tensioning

The next critical breakthrough for the design of post-tensioned column-supported floors was the extension of the strip method common for design of conventionally reinforced concrete to post-tensioned floors.

In the strip method, each floor is subdivided along the line of its supports into strips in two directions. Figure 1.16 shows the plan of a post-tensioned floor and its subdivision into orthogonal design strips. Each strip consists of a line of supports and its tributary. For structural design, the strips are considered in isolation extracted from the floor system. The extracted strip is analyzed and designed as a plane frame under the loading directly acting on the extracted strip when in place.

The strip method of design models the two-dimensional response of a column-supported floor as one-dimensional strips – somewhat similar to the familiar beam and one-way slab design.

1.6.3 Adoption of single 'representative' stress for design

The prototype post-tensioned floor responds to load as a plate. The stresses developed vary from point to point. High stresses at a 'point' on the plate lead to cracking. Unlike conventionally reinforced concrete, where cracking is anticipated and allowed, in post-tensioned floors the probability of cracks is to be predicted and accounted for.

The strip method does not reveal the distribution of stress across the design strip. Hence the location and value of crack-forming stresses are not given.

The shortcoming of the strip method in the prediction of the design condition for cracking is seen in the introduction of a single fictitious or representative design stress for each design section along the length of the design strip. This is explained next.

Figure 1.16 Breakdown of a floor slab in design strips.

Post-tensioning in buildings 15

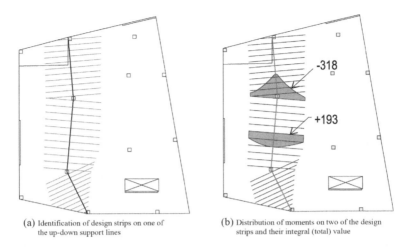

(a) Identification of design strips on one of the up-down support lines

(b) Distribution of moments on two of the design strips and their integral (total) value

Figure 1.17 Subdivision of design strips to design sections and extraction of design values.

Figure 1.17 (a) shows the plan of a column-supported floor along with design sections of a typical support line and its associated design strip. Part (b) of the figure shows the integral value of two of the design sections. One of the design sections selected in part (b) is at the face of the column support and the other at midspan. The design compliance with the building code for each of the design sections is checked using the integral value associated with the design section, in this case 193 for midspan.

To obtain the representative value for design, the integral (total value) is applied to the cross-sectional geometry of the design section to arrive at a single design value for code compliance.

The single stress is considered to represent the state of the entire design section for the prediction of probable crack formation. It is a representative stress, also referred to as 'computed' or 'fictitious' stress.

Both EC2 and ACI base the prediction of probable cracking in column-supported floors on the value of the 'fictitious' stress. Chapter 4 outlines the details for the computation of the fictitious stress using either EC2 or ACI 318.

In summary, the adoption of the design strip and single representative design value at each design section opened the opportunity for the design of post-tensioned column-supported floor slabs to practicing consulting engineers at large.

1.7 SOFTWARE APPLICATION

1.7.1 Strip method

Wide adoption of computers in the early 1980s by consulting engineers together with the application of simple load balancing paved the way to increased construction of post-tensioned building in the US. Computer software, such as 'ADAPT-PT[1]' based on the well-established strip method and 'extended load balancing' (Aalami, B. O., 1990) enabled the treatment of complex design strips with column drops, beams and other irregular features. Figure 1.18 illustrates an example of a computer model of a complex design strip.[2] The computer models faithfully reflect the geometrical features of the isolated design strip extracted from the floor system.

1.7.2 Single level finite element technology

Emergence of software based on the Finite Element Method (FEM) enabled floor slabs to be modeled and analyzed essentially in their true geometry. This bypassed the simplifications necessitated for geometry and tendon layout in converting the two-dimensional floor system to the one-dimensional traditional strip method.

Figure 1.19 illustrates the subdivision of the floor shown in Figure 1.16 (a) to finite element cells, its tendon layout and display of a solution example. Unlike the strip method that models the tendons in one direction only, the FEM-based design can feature the multi-dimensional layout of the tendons, each tendon represented with its unique profile and force.

The FEM solution of the modeled floor provides the computed contour of deflection as shown in part (c) of the figure, as well as distribution of other parameters such as moments, shears, local stresses and more.

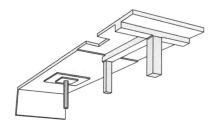

Figure 1.18 View of the computer model of a design strip using strip-based software.

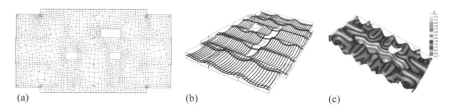

(a) (b) (c)

Figure 1.19 Design stages of a floor slab using finite element-based software.

Figure 1.20 Partial view of a post-tensioned floor using BIM.

While the finite element software has the capability to report the floor response in detail, such as stresses at a 'point' for crack prediction, the design procedure remains based on the strip method. Such design based on the strip method follows the current primary recommendations of EC2 and ACI 318 for column-supported floors. However, both codes also permit the application of FEM-based analysis and design as permissible procedures.

1.7.3 Multi-level and building information modeling (BIM)-based modeling and design

Modeling of post-tensioning tendons and the post-tensioning hardware is fully integrated in BIM-based software (Visicon, 2022). Beyond the traditional design features, the BIM models can detect tendon interference with other elements of construction. This eliminates job site trouble shooting. Figure 1.20 is a partial view of a BIM-generated post-tensioned floor showing the post-tensioning tendons and other inserts.

NOTES

1. www.RISA.com.
2. RISA-ADAPT-PT.

REFERENCES

Aalami, B. O. (1990), "Load Balancing – A Comprehensive Solution to Post-Tensioning," *ACI Structural Journal*, V. 87, No. 6, November/December 1990, pp. 662–670.

ACI 318-19 (2019), *Building Code Requirements for Structural Concrete (ACI 318-19) and Commentary*, American Concrete Institute, Farmington Hill, MI, www.concrete.org, 623 pp.

Billingtron, D. P. (2004), "Historical Perspective on Prestressed Concrete," *PCI Journal*, V. 49, No. 1, January–February 2004, pp. 14–30.

European Code EC2 (2004), *Eurocode 2: Design of Concrete Structures – Part 1-1 General Rules and Rules for Buildings*, European Standard EN 1992-1-1:2004, Brussels.

Kelley, G. S. (2003), "A Guide to the Components of an Unbonded Post-Tensioning System," *Concrete International, American Concrete Institute*, V. 25, No. 1, January 2003, pp. 71–77.

Lin, T. Y. (1963), "Load-Balancing Method for Design and Analysis of Prestressed Concrete Structures," *ACI Journal Proceedings*, V. 60, No. 6, June 1963, pp. 719–742.

Loewe, M. S., and Capekka-Llovera, J. (2014), "The Four Ages of Early Prestressed Concrete Structures," *PCI Journal*, V. 59, No. 4, Fall 2014, pp. 93–121.

Nawy, E. G. (1997), *Prestressed Concrete, A Fundamental Approach*, Prentice Hall, Upper Saddle River, NJ, 3rd ed., 938 pp.

Visicon (2022), *Software for BIM Visualization; Quantities; Error Detection*, www.PTstructures.com

Chapter 2

Application of post-tensioning in building construction

2.1 INTRODUCTION OF POST-TENSIONING IN BUILDING CONSTRUCTION

Post-tensioning in common building construction was first introduced in the US in the early 1950s. The objective was to eliminate cracks and reduce deflection in thin flat slabs.

The principal design instrument for its application was put forward by T. Y. Lin (1963) through the concept of load balancing. In its basic form, load balancing allows the engineer to view the effects of post-tensioning in reducing the design dead load of the slab. The simple load balancing concept was expanded to cover the design of nonprismatic members and general configurations (Aalami, B. O., 1990, 2007).

The post-tensioning hardware of the mid-1950s evolved through several significant improvements to a durable and corrosion-resisting item (Aalami, B. O., et al., 2016). The improvement in hardware resulted in extended durable building construction. The following is a selection of the current applications of post-tensioning in residential and commercial building construction.

Figure 2.1 is an example of a post-tensioned multi-story flat slab construction.

2.2 COLUMN-SUPPORTED FLOORS

Modern column-supported floors are typically flat plates essentially without beams or column drops/caps or drop panels.

2.2.1 Flat slab construction

Figure 2.2 illustrates the interior view of a typical beamless flat slab construction. It provides a clear and continuous full-height ceiling between the core wall and the building's exterior.

DOI: 10.1201/9781003310297-2

Figure 2.1 W-Hotel, courtesy CKC.

Figure 2.2 Example of a post-tensioned flat slab construction.

2.2.2 Podium slabs

Where commonly four to six levels of residential or office space rest over parking or retail space, the optimim space allocation between the parking/retail below and the residential/office above favors change in the layout of the vertical supports.

Retail space benefits from large column spacing. Parking requires specific subdivision of space that dictates the column locations. For this reason, it is not uncommon, specifically in the US, to construct the parking/retail space using supports that are wide apart and post-tensioned floors. The uppermost post-tensioned floor – referred to as a podium slab – is designed to support several levels of light-framed super structure that is typically for residential or office occupancy and has a different wall/column layout than the levels below.

Figure 2.3 is an example of hybrid construction with a podium slab supporting light framing construction.

2.3 BEAM AND SLAB CONSTRUCTION

When the aspect ratio of a panel is 2 or more, it is more economical to use beams along the long direction and flat slab between the beams. This is a preferred configuration for parking structures with ramps accommodating parking stalls on each side and two-way traffic in the middle.

Figure 2.4 is the view of a post-tensioned beam and slab parking structure construction. Typically the beams span 18–20 m and are spaced 5 m apart.

Figure 2.3 The uppermost post-tensioned flat slab podium supports several levels of light framing construction.

Figure 2.4 View of a typical parking structure framing.

2.4 GROUND-SUPPORTED CONSTRUCTION

By far the largest volume of post-tensioning in the US is in ground-supported slabs in three different applications. These are mat foundations, slabs on expansive soil and industrial slabs.

2.4.1 Mat foundation

Mat, also referred to as raft foundation, is used mostly in moderate height buildings supported on soil with low allowable bearing pressure or likelihood of large differential settlement. Post-tensioning in a contiguous foundation slab can be configured to distribute more uniformly the column and wall loads from above over the foundation soil below. Figure 2.5 is the schematic of the basics of the post-tensioned mat foundation.

Figure 2.6 is the close-up of a post-tensioned mat foundation. It shows the tendons at the low point below the column, arranged such as to counteract the downward force of the column on the foundation.

2.4.2 Slabs on expansive soil

Deformation of light building structures, typically of one to three levels, or light industrial buildings on shallow foundation is sensitive to changes in the level of the supporting soil.

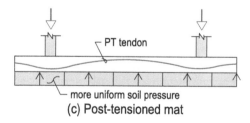

Figure 2.5 Schematic of the basics of a post-tensioned mat foundation.

Figure 2.6 Post-tensioned mat foundations.

Seasonal changes in soil moisture result in volume change of expansive foundation soil. This can result in the deformation in the soil-supported light building to the extent that it exceeds the building's serviceability limits.

Post-tensioning in ground-supported slabs on expansive soil can be designed to reduce the deformations experienced by the supported structure.

The design objective of a post-tensioned foundation slab is to ride over the changes in elevation of the soil support. Figure 2.7 shows an example of a post-tensioned foundation slab on expansive soil ready to receive concrete.

2.4.3 Industrial floors

Where the foundation slab is required to be contiguous, flat, crack-free and jointless for industrial applications, in particular warehousing, post-tensioning is a suitable option to meet the challenge. Figure 2.8 shows an example of a post-tensioned industrial ground-supported slab.

The function of post-tensioning in industrial ground-support slabs is to:

(i) Avoid the formation of shrinkage cracks, or close the cracks at stressing.
(ii) Provide precompression to reduce the likelihood of cracking due to concentrated loads from storage racks or forklifts.

Figure 2.7 Post-tensioned building foundation slab.

Figure 2.8 View of industrial slabs under construction.

2.5 TRANSFER FLOORS

Change in the continuity of vertical supports through a floor requires the latter to transfer the load of the discontinued columns/walls above to the supports below. Figure 2.9 illustrates the typical application of a transfer plate. Hotel lobbies, shopping malls and casinos below highrise residential/hotel construction are common examples.

Depending on the arrangement of the structure it is not uncommon for transfer floors to be as thick as 3–4 m.

Figure 2.10 is the construction view of a 1.5 m transfer plate for an apartment house ready to receive concrete. The floor is more than 1 m deep. It is reinforced with bonded tendons.

The application of post-tensioning helps to transfer the somewhat closely located concentrated loads on the plate to the few supports below.

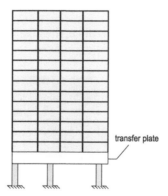

Figure 2.9 Example of typical arrangement transfer plate.

Figure 2.10 Transfer plate under construction.

2.6 REPAIR AND RETROFIT

Post-tensioning has been used to correct deflection and add strength to existing structures. Figure 2.11 illustrates an example for strength and deformation retrofit of a member through application of external post-tensioning.

The concept of external post-tensioning was used successfully in the rehabilitation of the Pier 39 parking structure in San Francisco (Aalami, B. O., and Swanson, D. T., 1988). Figure 2.12 shows the installaton of external post-tensioning on the side of the beam web. When in service, the underdesigned structure developed excessive deflection and cracking. Application of external post-tensioning restored the integrity of the structure.

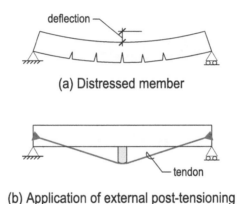

Figure 2.11 External post-tensioning to close cracks and add strength.

Figure 2.12 External tendons are installed one on each side of the beam's web.

2.7 APPLICATION IN HIGH-RISK SEISMIC ZONES

2.7.1 Flat slab construction

In high-risk seismic zones, such as San Francisco, thin post-tensioned flat slabs prove to be the choice for multi-level concrete construction. The slabs are designed and detailed to act as membranes between the designated seismic load-resisting members of the concrete framing. The slabs are detailed to be ductile at connections to the lateral force resisting members of the structure. This is to accommodate the seismic-induced horizontal displacement and rotations while the slabs and columns retain their integrity to resist the gravity loads. Figure 2.13 illustrates the identification of the lateral force resisting members of the framing and their connectivity.

2.7.2 Correction of seismic deformation

In high-risk seismic regions, such as much of California, buildings are designed to undergo post-elastic deformation when subjected to design seismic force. Local post-elastic response helps to reduce the seismic force on the structure.

The buildings are designed to prevent collapse under anticipated seismic forces, but damage is expected.

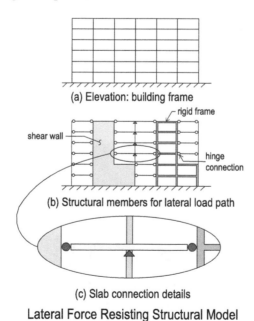

Figure 2.13 Identification of the lateral force resisting members of the framing and their connectivity.

Observations of past earthquake damage in California revealed that several multi-story buildings did not fully return to their plumb position after the earthquake motion. The out-of-plumb position can lead to operational and maintenance problems for the building.

Figure 2.14 illustrates the phenomenon by way of the idealized seismic force/horizontal displacement of a building undergoing post-elastic response. At termination of the force (Force = 0), such as point A on the graph, the frame with post-elastic response is likely to have the finite displacement (d).

Post-tensioning can be used to help in restoring the building frame closer to its original position. This is achieved by directing and controlling the post-elastic deformation of the frame to designated locations, and using the force of prestressing tendons to help to restore the building to its original plumb condition.

Figure 2.15 explains the concept for the displacement correction mechanism. Consider three members assembled to form the frame. The connection is secured by an elastic through an oversized hole (part a).

The horizontal force F displaces the frame as shown in part (b) of the figure. The applied force is counteracted by the extension and force increase in the elastic element. At removal of force F, the tension increase in the elastic element acts to restore the frame to its original geometry.

The interface between the beam and the column opens and closes as the force F varies (part c).

As an additional measure, the stretching and contraction of inserted elements across the interface between the beam end and column face can be configured to absorb part of the energy from the applied force. This reduces the frame's horizontal displacement.

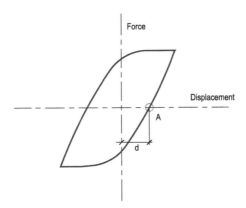

Force-Displacement in Post-Elastic Regime

Figure 2.14 Idealized lateral force–displacement relationship of a building frame undergoing post-elastic response.

28 Post-tensioning in building construction

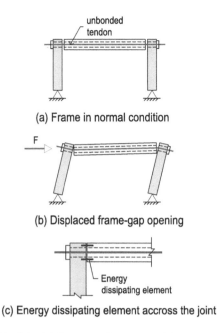

Figure 2.15 Model of frame with displacement correction.

Figure 2.16 shows a building in San Francisco that is constructed using the seismic deformation correction concept (Englekirk, R. E., 2002).

The building features precast concrete frames designated to help recover the deformation of the building after an earthquake.

Figure 2.17 illustrates the construction details and the function of the deformation restoring assembly.

Figure 2.18 shows a seismic displacement correction frame under construction (part a) and when installed (part b).

2.8 REDISTRIBUTION OF REACTIONS

Post-tensioning is commonly used to reduce deflection, control cracking and add strength in a wide range of construction projects, including both new construction and retrofit of existing structures. The principal features of post-tensioning used in common design are the precompression that is applied to the concrete and the uplift that is generated to offset gravity loads. A third characteristic of post-tensioning is the generation of hyperstatic (secondary) forces in statically indeterminate structures. Hyperstatic

Application of post-tensioning in building construction   29

Figure 2.16 Paramount building, with seismic deformation control feature (San Francisco, California).

forces can act at the primary design parameter in special applications. An example is explained next.

Hyperstatic forces from post-tensioning were used to resolve a major challenge facing the structural design of a high-rise in New York City that is partially constructed over, and supported by an existing adjoining structure (Figure 2.19 is a view of 55 Hudson Yards, where a portion of it rests on the adjacent construction). The design scheme required the columns of the existing bottom structure to support the new construction overhang above it.

The engineering challenge was to match the reactions of the new construction to the location and capacity of the supporting columns of the existing structure below.

While the new structure was designed in total weight to match the total capacity of the existing four supports, the distribution of the overhang reactions was widely different from the capacities of the individual supports (Aalami, B. O., et al., 2016).

Post-tensioning installed in the walls of the new construction was configured to generate a set of hyperstatic reactions so that the combined gravity and hyperstatic reactions from the new structure matched the capacity of the existing supports.

Figure 2.20 describes the concept for generation of support reactions through post-tensioning. When a tendon container cannot freely deform under the forces from post-tensioning (part c), hyperstatic forces develop at the supports. The zero sum hyperstatic forces do not change the sum of

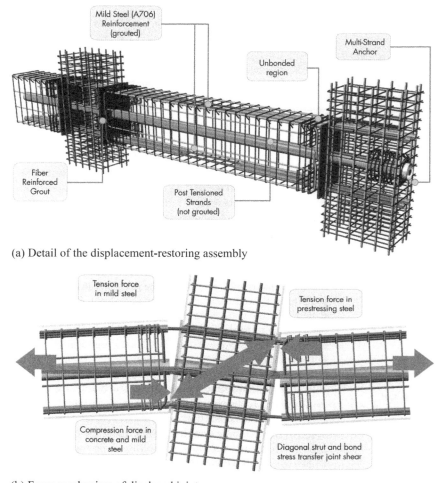

(a) Detail of the displacement-restoring assembly

(b) Force mechanism of displaced joint

Figure 2.17 Arrangement of post-tensioning and reinforcement at the seismic restoration joint. Courtesy Suzanne Nakaki; CI 2017.

reactions of the member from other causes, but they modify their magnitudes individually.

Figure 2.21 shows the arrangement of tendons in the walls of the multistory building to correct the distribution of the building's reactions. In Figure 2.21 W is the reaction of the building. It is the function of the building's geometry and material. H is the hyperstatic force at the foundation from the post-tensioning tendons. The hyperstatic forces add up to zero. R is the reaction force delivered to the foundation. It conforms with the foundation capacity.

Application of post-tensioning in building construction 31

(a) Precast frame detailed for seismic deformation recovery

(b) Installed seismic deformation recovery frame

Figure 2.18 Building with seismic deformation recovery feature in Roseville, California.

Figure 2.19 Hudson Yards, New York with support reaction control feature.

32 Post-tensioning in building construction

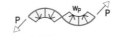

(a) $\sum$ Force=0
Tendons are in self-equilibrium

(b) When constrained in a container,
tendons change the shape of the container

(c) When container is not free to change its shape,
forces develop at constraints.
The constraint forces add up to zero.

Figure 2.20 Development of hyperstatic forces from post-tensioning.

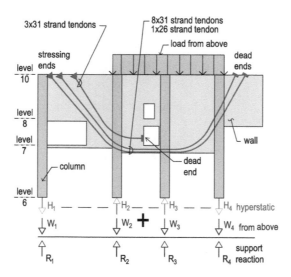

Elevation : Tendon Arrangement in Wall

Figure 2.21 Wall elevation at levels 7 through 10 showing the arrangement of tendons.

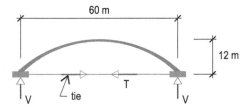

Figure 2.22 Elevation: structural system of a long-span sports hall.

2.9 SPECIAL APPLICATION OF POST-TENSIONING

Long-span exhibition halls, tension structures and shell structures often feature tensioned cables in one form or another to resist high-tensile forces.

Figure 2.22 shows the schematics of a tension tendon resisting the horizontal forces of a long-span sports hall. Similar arrangements are often used in other exhibition or sports halls, such as Moscone Exhibition Hall in San Francisco.

2.10 WALLS AND FRAMES

Post-tensioning has been used in special applications of walls and frames both for gravity and seismic force designs (Englekirk, R. E., 2002; Stevenson, M., et al., 2008).

REFERENCES

Aalami, B. O. (2007), "Critical Milestones in Development of Post-Tensioned Buildings," *ACI, Concrete International*, V. 29, October 2007, pp. 52–52.

Aalami, B. O. (1990), "Load Balancing – A Comprehensive Solution to Post-Tensioning," *ACI Structural Journal*, V. 87, No. 6, November/December 1990, pp. 662–670.

Aalami, B. O., Aalami, F. B., Smilow, J., and Rahimian, A. (2016), "A Novel Application of Post-Tensioning Solves High-Rise Design Challenges," *Concrete International*, V. 38, October 2016, pp. 58–63.

Aalami, B. O., and Swanson, D. T. (1988), "Innovative Rehabilitation of a Parking Structure," *American Concrete Institute, Journal of Concrete International*, V. 10, No. 2, February 1988, pp. 30–35.

Englekirk, R. E. (2002), "Design-Construction of the Paramount – A 39-Story Precast Prestressed Concrete Apartment Building," *PCI Journal*, V. 47, No. 4, July–August 2002, pp. 56–71.

Farid, E. (2017), "Integrating Precast Cladding and Structure," *Concrete International*, V. 39, No. 10, October 2017, pp. 38–41.

Lin, T. Y. (1963), "Load-Balancing Method for Design and Analysis of Prestressed Concrete Structures," *ACI Journal Proceedings*, V. 60, No. 6, June 1963, pp. 719–742.

Stevenson, M., Panian, L., Korolyk, M., and Mar, D. (2008), "Post-Tensioned Concrete Walls and Frames for Seismic Resistance – A Case Study of David Brower Center," *Structural Engineering Association of California Convention Proceedings*, pp. 1–8.

Chapter 3
Basics of post-tensioning design

3.1 PRESTRESSING STEEL

The effective application of post-tensioning is largely due to markedly higher tensile strength of prestressing steel compared to conventional reinforcement. Figure 3.1 highlights the relative strength of the two materials.

The modulus of elasticity of mild steel and prestressing steel is essentially the same. But their yield stress is widely different. The post-tensioning steel must be stretched typically 3.5–4 times more than common rebar before it yields and develops its full strength. The stretching necessary to yield prestressing steel embedded in concrete as rebar results in unacceptable concrete cracks.

Post-tensioning provides the means of taking advantage of the high strength of prestressing steel as reinforcement while avoiding unacceptable cracking in concrete. Post-tensioning imparts precompression in the concrete that is prestressed.

Once in service, the applied load must first overcome the initial precompression from post-tensioning before concrete develops tension. For this reason, precompression from prestressing enables the effective use of higher strength prestressing steel in concrete structures.

3.2 POST-TENSIONING DESIGN REQUIREMENTS

Like other structural elements, post-tensioned members are designed for service condition, namely the serviceability limit state (SLS) and safety condition, referred to as the ultimate limit state (ULS). In addition, at time of application of post-tensioning – jacking – a post-tensioned member will be subject to forces that may compromise the member's resistance. Hence, post-tensioned members are also checked for 'Initial Condition' in addition to serviceability and safety limit states.

36 Post-tensioning in building construction

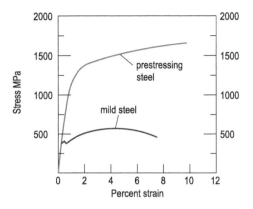

Stress - Strain Diagrams

Figure 3.1 Stress-strain diagram of typical prestressing steel and common reinforcement.

The primary load combinations for the serviceability (SLS) and safety (ULS) are:

Serviceability (SLS)

$$U = 1.0DL + 1.0LL + 1.0PT \tag{3.1}$$

Safety (ULS)

$$U = K_D DL + K_L LL + K_{PT}(PT \text{ or } HYP) \tag{3.2}$$

where

K_D, K_L are dead and live load factors. These are defined in the building code used for design of the member. Chapter 4 gives the values for EC2 and ACI 318 building codes.

K_{PT} is the factor for either post-tensioning (PT) or the hyperstatic effects from post-tensioning (HYP).

Whether PT or HYP is used for safety checks depends on the method of design. The following details the options.

3.3 POST-TENSIONING DESIGN METHOD OPTIONS

There are three options to design a post-tensioned member. The following explains the basics and the differentiating features of each of the options.

Figure 3.2 shows a member with post-tensioning and nonprestressed (rebar) reinforcement. Review the force condition of the member on its own

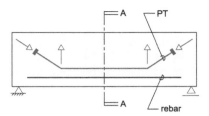

(a) Post - tensioned member
showing PT force on concrete

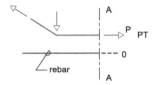

(b) Assumed initial force is
PT and reinforcement

Initial Post - Tensioning
and Rebar Forces

Figure 3.2 Post-tensioned member with conventional reinforcement.

prior to the consideration of self-weight or other loads. In the initial condition, the tension in the post-tensioned tendon results in application of forces in the member shown with the arrows.

The forces shown by the arrows in part (a) of the figure result in deformation of the member and some force in the rebar. For design, it is commonly acceptable to assume that at a typical section such as A-A in part (b) of the figure the initial force in the rebar is zero, or small.

Depending on how the force *P* in the post-tensioned tendon and the force in rebar – shown as zero – are treated in the serviceability and safety stages of the design, three different design procedures are available. These are referred to as:

- Straight method.
- Load balancing method.
- Comprehensive method.

All three methods meet the EC2 and ACI code requirements, but differ in the design effort and possibly the economy of design.

Figure 3.3 shows a segment of rebar and post-tensioning tendon from the service and safety conditions of the member. The reinforcement segment shown in Figure 3.3 (a) will be used to describe the options.

Part (ii) of the figure shows the force P in a typical post-tensioned tendon. This is the force that the tendon exerts on the member in service condition. In the following discussion, this is referred to as PT force.

The PT forces acting on the member are in static equilibrium. Their sum adds up to zero.

The PT forces tend to change the member's shape. If the member supports restrict the member's free deformation, restraining forces (reactions) develop at the member supports. The support reactions are referred to as 'Hyperstatic HYP' forces from post-tensioning. The support reactions add up to zero, since the sum of the forces that causes them, namely post-tensioning forces, is zero.

Two of the three design options, namely the 'straight method' and 'load balancing method,' differ in the selection of either the PT or HYP forces in the evaluation of the member's safety condition (ULS). The following explains the underlying difference.

Refer to Figure 3.3 (a). Under service condition, the rebar in the member is 'assumed' to be under zero stress. The force in post-tensioning tendon is P. This is the effective force after immediate and long-term losses. It is identified as PT in the preceding load combinations.

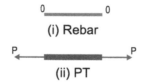

(a) Service condition (SLS)

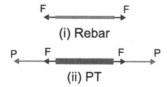

(b) Strength condition (ULS)

Response of Rebar and PT

Figure 3.3 Stress condition of rebar and PT in service and safety condition.

The force assumption zero for rebar and P for PT shown in part (a) of the figure governs the service condition of the member, specifically the member's deformation and stresses. In the common design practice, the force in rebar – if any – that may arise from deformation of the member under service load is disregarded. The force in rebar is assumed zero; the force in the PT tendon is assumed P. Again, P is the long-term effective service force after all losses.

For safety evaluation of the member, the contribution of rebar is considered. Often rebar is required and sized to supplement the post-tensioning in meeting the safety requirements of the member.

Refer to part (b) of the figure, where at ULS the rebar is assumed stretched to its yield point. The force in a typical rebar is shown by F. The force F is based on the extension of the rebar at ULS and the rebar's modulus of elasticity.

A bonded tendon adjacent to the rebar will undergo the same extension as rebar. Having essentially the same modulus of elasticity, the initial force P in the bonded tendon will increase by its area times the same stress increase as in rebar, namely F for the same area. In effect, the stress in tendon at ULS will be that due to the service condition force P plus an increase similar to the adjacent rebar, namely $(P + F)$ for the same PT and rebar area.

It is the recognition and treatment of the force F at the ULS in the prestressing steel that leads to three different design schemes. The methods differ by whether or not to account for the stress increase in the tendon associated with F in an adjacent rebar, and to what extent to account for it. Again, there are three options:

- Straight method.
- Load balancing method.
- Comprehensive method.

First, a word on design significance of the force stress F at ULS of common building construction. EC2 and ACI 318 both recognize and allow for an increase in tendon force at ULS. But they differ largely on how to account for the increase. This leads to preference in design method and selection of a post-tensioning system based on the applicable code.

A. EC2 provision

Consider a member reinforced with industry common bonded tendons. At ULS the maximum design contribution of P to the resisting moment is f_{pk} = 1,860 MPa. With the effective stress in SLS being typically 1,000 MPa for bonded tendons, the increase in stress at ULS can be as much as 760 MPa, namely about a 40% increase in contribution from post-tensioning compared to SLS.

The 40% increase in contribution of prestressing when using bonded tendons is significant enough to justify the complexity of a design method that accounts for it. This is specifically valid if post-tensioning is a significant component of the construction scheme.

Next, consider the same member reinforced with unbonded tendons. Based on EC2, at ULS the maximum increase in tendon force F is based on 100 MPa stress.[1] When compared to the effective long-term stress 1,100 MPa of strand used in service condition (SLS), the increase in tendon contribution at ULS is 9%.

The economy for the selection of bonded or unbonded tendon in design clearly favors the bonded system since the gain in stress of bonded tendons is about four times compared to the unbonded alternative. For the same amount of post-tensioning the bonded system contributes more.

B. ACI 318 provision

For members reinforced with bonded tendons the increase in strand stress at ULS is permitted to be as high as 414 MPa. This is an increase of 414/1,860 = 22% in the strand's force contribution.

For unbonded tendons, the maximum allowable increase in tendon stress at ULS is 207 MPa.[2] This translates to 207/1,860 = 11% effectiveness.

The larger allowable stress increase for unbonded tendons in ACI 318 at ULS, compared to EC2, tends to favor the selection of unbonded tendons for design.

3.4 STRAIGHT METHOD

The straight method is the simplest of the three design options. It accounts for the same contribution of prestressing in service condition as that of the load balancing method. This will be explained in the next section. For the safety condition, however, the contribution of post-tensioning is handled differently.

The straight method does not require the hyperstatic actions from prestressing to be computed and accounted for as an *explicit* design consideration. The inclusion of hyperstatic actions is *implicit* in the design process. This is advantageous for engineers who do not deal with post-tensioning on a regular basis, and may not be conversant with the computation of hyperstatic actions.

The load combination for the 'total' service condition is the same as the other two design methods. As an example, the 'total' or 'characteristic' load combination is:

$$U^3 = 1.00DL + 1.00LL + 1.00PT \tag{3.3}$$

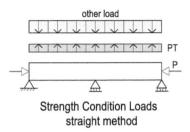

Figure 3.4 Ultimate limit state (ULS) load diagram using the straight method.

For the safety step of the design (ULS), post-tensioning is viewed as an externally applied load, similar to the service condition. Unlike the other two methods, the post-tensioning tendons are not considered as reinforcement for providing resistance to the applied load – rather they are viewed as *applied load*. The entire required resistance to the 'computed' design moment, if any, is provided by non-prestressed reinforcement, along with compression from prestressing.

Figure 3.4 illustrates the safety (ULS) condition loads on the post-tensioned member. For this condition, there is no post-tensioning tendon in the member. But the effect of post-tensioning is accounted for through the load it applies on the member.

In summary, at ULS the contribution of prestressing is accounted for in (i) reduction of demand moment, and (ii) precompression in enhancing the resistance capacity of the member.

For the straight method of design, the load combination for the ULS using EC2 is:

$$U = 1.35DL + 1.5LL + 1.00PT \tag{3.4}$$

Note that in the above load combination, HYP (hyperstatic) is replaced by PT, when compared with the other two methods.

The preceding method is complete, valid and code-compliant in arriving at safe designs. It includes the presence of hyperstatic effects with load factor 1.00. The straight method is not the most economical alternative, however. Its simplicity and expediency justify its application, where economy of the member is not of critical concern.

It is important to recognize that the term PT in the load combination includes the hyperstatic actions (HYP) as required both by EC2 and ACI 318.

3.5 LOAD BALANCING METHOD

Load balancing relies on basic knowledge of structural engineering, as opposed to new skills and unfamiliar concepts. This is true except for the recognition and explicit computation of hyperstatic moments from post-tensioning. The calculation of the hyperstatic moments is necessary for the safety compliance of the member at ULS. The understanding and treatment of the hyperstatic effects is generally an obstacle for many structural engineers.

Figure 3.5 shows the impact of post-tensioning in reduction of the effects of other loads on the member. The figure is used to explain the concept in its basic form. For illustration of the concept, the tendon shown in part (a) is in the shape of a simple parabola in each of the spans. Pulled to force P, the tendon is approximated to exert a uniform uplift (w_p) shown in part (b). The uplift is considered to reduce the effects of the other loads of the structure (Q), to (Q-w_p).

For deflection, stresses and crack control, which are part of the serviceability check of the member, the effective downward force on the member is reduced by w_p. The reduced downward force improves the service performance of the member.

The EC2 characteristic service load combination for 'total load' is given in Expression 3.5. The same load combination is used for ACI 318 (total) condition:

$$U = 1.00DL + 1.00PT + 1.00LL \qquad (3.5)$$

At ULS, the member relies on the contribution of the tendons as reinforcement in resisting the demand forces.

Hence the post-tensioning tendon is viewed back in the member. Placing the tendon back in the member to provide resistance, the configuration and

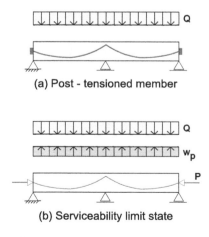

Figure 3.5 Contribution of post-tensioning in reduction of the effect of dead load on member.

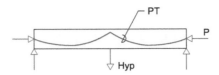

PT Reactions at Ultimate Limit State

Figure 3.6 Hyperstatic reactions from prestressing.

load condition shown in Figure 3.6 applies. Figure 3.6 illustrates the generation of support reactions (hyperstatic forces) resulting from the action of the post-tensioning tendon on the statically indeterminate member.

The member configuration shown in Figure 3.5 (b) – with missing tendons – would not qualify. With the tendons back in the member, the configuration Figure 3.6 applies.

Flexing of the member under the tendon forces, coupled with the restraint of the supports to the member's unrestricted change in shape, result in the reactions (HYP) at the supports (Figure 3.6). It is reiterated that the post-tensioning reactions are generated when the member is not free to flex – hence the term 'hyperstatic,' also referred to as secondary reactions.

The hyperstatic reactions (R), in turn, result in forces, such as moments and shears, in the member that must be resisted by the member reinforcement that includes post-tensioning tendons. Reinforcement is added if post-tensioning tendons are not adequate to resist the demand moment.

For the safety of the structure, at ULS, EC2's primary load combination is:

$$U = 1.35DL + 1.5LL + 1.0HYP \tag{3.6}$$

The ACI 318 load combination is:

$$U = 1.20DL + 1.6LL + 1.00HYP \tag{3.7}$$

where HYP is the actions from prestressing resulting from the hyperstatic reactions at the supports.

The resistance to the force demand is provided by concrete and prestressing tendons. At ULS the force in prestressing tendons is increased by F described in Section 3.3. The shortfall in resistance, if any, is met by the addition of non-prestressed reinforcement.

In short, in the load balancing method (i) the tendon is viewed removed to check the service condition; (ii) the tendon is in place acting as reinforcement to check the safety condition.

Simple load balancing was introduced by T. Y. Lin in 1963. It accounted for the bending effects on the member resulting from the lateral forces from post-tensioning. Its application was essentially limited to prismatic members with common tendon arrangements.

Load balancing was extended by Aalami, B. O. (1990) to include the effects of axial forces from post-tensioning on bending effects of non-prismatic members. The extension made it feasible to design members with changes in cross-sectional geometry and general tendon layout.

3.5.1 Simple load balancing

Figure 3.7 (a) shows a simply supported member of uniform cross-section under the externally applied dead (DL), and live load (LL). A tendon in the shape of a simple parabola provides the post-tensioning. The tendon is anchored at the centroid of the member at each end.

Part (b-i) of the figure shows the member without its post-tensioning tendon. Part (b-ii) shows the forces that the post-tensioning tendon exerted on the member when the tendon was in place. To restore the balance of forces

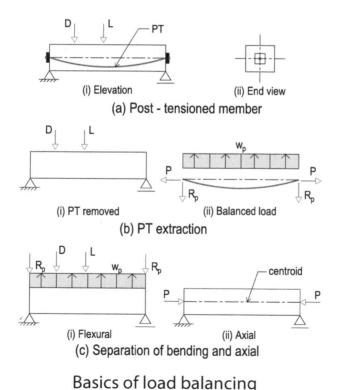

Figure 3.7 Basics of balanced loading.

in the member from which the tendon is removed, the tendon forces shown in part (b-ii) of the figure must be placed back in the member.

The tendon forces shown in part (b-ii) of the figure consist of up and down forces (w_p, R_p), and axial force P. In simple load balancing, it is assumed that the forces P at the ends of the member are equal and collinear. They balance one another. More importantly, they do not cause bending of the member. Hence, for analysis of bending effects, the axial forcer P can be considered separately from the remainder of the post-tensioning forces (parts b-ii and c-ii).

The impact of the force P for a prismatic member with the tendon anchored at the centroid of the member is simply uniform precompression (*f*). The value of the compressive stress *f* is given by:

$$f = P/A \tag{3.8}$$

The remainder of the post-tensioning forces w_p (up and down forces from tendon) and the vertical reactions of the tendon (R_p) are considered back on the member (Figure 3.7 (c-i)).

In this case, the substitution of post-tensioning by an equivalent force enables the structure to be treated like a non-prestressed member under bending.

3.5.2 Extended load balancing

The central premise of the simple load balancing method, as outlined above, is decoupling of the bending and axial effects. The axial effects are assumed to cause uniform compression in the member. For members of uniform cross-section with tendons anchored at the centroidal axis of the member, this can be achieved by simply substituting a tendon only by its lateral forces for bending effects of the member.

Figure 3.8 shows the balanced loads in a post-tensioned member with shift in centroidal axis. Part (a) shows the tendon anchored at the centroid of the member at each end. The change in member thickness results in a shift in the location of the centroidal axis between the two spans.

Part (b) of the figure shows the tendon removed from the member. Once removed, the forces that the tendon exerted, when in place, are shown in the figure. These include vertical forces along the member and axial forces at the member ends.

The shift in the centroidal axis (*m*) results in moment M_p. This moment adds to the up and down forces from post-tensioning for the bending analysis of the member.

The shift between the line of actions of the forces at the ends of the member results in bending moment M_p in the member. This violates the premise of uniform compression from prestressing tendons required in simple load balancing.

46 Post-tensioning in building construction

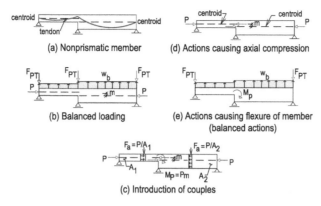

Balanced Loading and Change in Member Centroid

Figure 3.8 Balanced load in post-tensioned member with shift in centroidal axis.

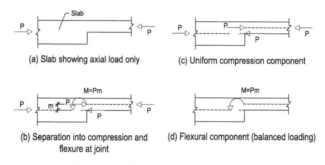

Treatment of Non-congruent Axial Forces

Figure 3.9 Treatment of non-aligned axial forces in post-tensioned member.

The premise of uniform compression can be restored by envisaging a clockwise and a counterclockwise couple each equal to Pm that act at the change in the centroid (part c) of the member. This addition does not disturb the equilibrium of the system.

One of the added couples can be considered to act with the axial loads of part (d) of the figure to restore the premise of uniform compression in each segment of the member. The second couple shown in part (e) is added to the flexural loads from the tendon.

Figure 3.9 illustrates the treatment of non-aligned axial forces in post-tensioned members. The force arrangement shown in part (c) will be used for the computation of axial effects. The moment in part (d) will act with other loads for bending effects.

3.5.3 Load balancing summary

Figure 3.10 illustrates the elements of the load-balancing design scheme. Part (b-i) shows the forces on the member exerted by the post-tensioning using simple load balancing. Part (c) of the figure recognizes the non-prismatic configuration of the member by the addition of a moment at change in the member's centroidal axis. The combination of the post-tensioning forces shown in parts (b) and (c) results in the support reactions identified in part (d) of the figure.

For serviceability design of the member, specifically deflection and stresses, the entire forces shown in parts (b, c, d) are accounted for. For the safety check of the member the hyperstatic forces identified in part (d) are considered.

The computation of effects of hyperstatic action on the member, such as moment and shear, are commonly carried out using one of two schemes, namely the *direct* and *indirect* methods.

In the direct method, first the support reactions are calculated. Next, using the support reactions, the hyperstatic values of the moment and shear along the member are worked out. These calculated hyperstatic forces are required to be included in the safety load combination. An example of this procedure is given in Chapter 6.

The *indirect* method works well for prismatic members and simple tendon layout (Aalami, B. O., 1990).

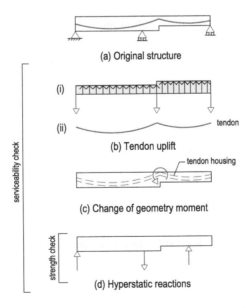

Figure 3.10 Actions from load balancing and their participation in design process when using load balancing method.

In the indirect method the hyperstatic moments are given by:

$$M_{hyp} = M_p - P \times e \tag{3.9}$$

where

e = eccentricity of post-tensioning/prestressing with respect to the neutral axis of the section at which the hyperstatic moment is sought.
M_{hyp} = hyperstatic moment.
M_p = post-tensioning balanced moment due to balanced loads.
P = post-tensioning/prestressing force.

3.6 COMPREHENSIVE METHOD

Unlike load balancing, where the post-tensioning tendon is considered as an applied load for service condition, in the comprehensive method the tendon is viewed as reinforcement with an initial stress (Aalami, B. O., 2000).

The primary application of the comprehensive method is where a member's deformation during construction or in service is critical to the performance of the structure. Segmentally constructed bridges, such as balanced cantilevers and cable-stay bridges, are typical examples for application of the comprehensive method.

Figure 3.11 highlights the features of the comprehensive method and its comparison with the load balancing. The sub-images (b, d, f) on the left of the figure represent the load balancing method applied to a segment of the member. For the segment identified, the prestressing is considered as applied load. The sub-images (c, e, g) on the right represent the comprehensive method, where the prestressing steel is viewed as reinforcement with initial stress.

In the load balancing sub-figures the prestressing tendon is removed. It is replaced by the forces it exerted, when in place (parts b and d). At application of load, and lapse of time, the member deforms (part f). The stresses in the deformed condition are changed due to losses in prestressing, and change in dimensions of the segment from creep, shrinkage and other stresses. The design method does not account for the change in member deformation resulting from time-dependent effects and other stresses. The changes in member deformation caused by time-dependent effects must be calculated and accounted for separately.

In the sub-figures on the right that represent the 'comprehensive' method the tendons are initialized with forces at stressing (P). Non-prestressed base reinforcement, if any, is initialized with zero stress.

Application of load, lapse of time, and due allowance for loss of stress in prestressing from creep, shrinkage of concrete and relaxation in prestressing lead to the solution shown in part (g). The solution applies to a given

Basics of post-tensioning design 49

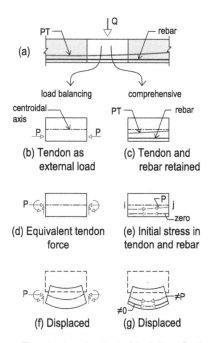

Post-Tensioning Analysis Modeling Options

Figure 3.11 Comparison between load balancing and comprehensive modeling methods.

time and load. It includes the time-dependent effects on prestressing and concrete. The necessity of post-solution computation of long-term losses and their allowance – as is the case in the load balancing method – does not arise.

The breakdown of the solution at any stage into contributions from dead load, live load, prestressing, creep and shrinkage provides the necessary information for the serviceability and safety load combinations of the member.

The necessity of computing hyperstatic actions from prestressing and including them in the ULS load combination as an item remains, as in the case of load balancing.

The 'total' load combination for service condition is:

$$U = 1.00DL^* + 1.00LL \qquad (3.10)$$

Note that in this case, post-tensioning is an integral part of the structure similar to concrete. DL^* includes the effects of post-tensioning as well as the presence of non-prestressed steel. For this reason, PT or HYP do not appear explicitly in the load combination.

For safety of the member, both EC2 and ACI 318 consider different load factors for the dead load of the structure and the effects of prestressing. The code requirement of different load factors for the *DL* and *PT* effects necessitates that for the safety check of the member the contribution of prestressing be determined separately.

The common EC2 load combination is:

$$U = 1.20DL + 1.60LL + 1.00HYP \tag{3.11}$$

Note that *DL* and *HYP* have different load factors. The same is true for ACI 318. Hence the necessity to extract the effects of post-tensioning. This step adds a level of complexity to the designs based on the comprehensive method.

3.7 APPLICATION OF THE METHODS

Three features enhance the performance, and improve the economy of a post-tensioned member. These are: (i) uplift, (ii) precompression, and (iii) gain in tendon stress at ULS.

Two of the design methods, namely *load balancing* and *comprehensive* method take advantage of all three features.

The *straight* method in its simplest form accounts for the benefits of the tendon uplift only. Allowing for precompression at ULS improves the economy of this design option. The method does not benefit from the gain in tendon stress at ULS, however.

The straight method is safe and expeditious. It eliminates the effort of computing the hyperstatic actions from prestressing as a separate design item, and bypasses its explicit inclusion in design.

At ULS, prestressing steel is stretched beyond its service condition. The stress gain from increased tendon stretching, among other factors, depends on whether the tendons are bonded or unbonded.[4] Using ACI-318,[5] the stress gain for bonded tendons can be as much as 414 MPa, and for unbonded tendons up to 207 MPa. Assuming rebar at 400 MPa, the gain in tendon stress translates to half the cross-sectional area of unbonded tendons and equal to the full cross-sectional area of bonded tendons contributed as added rebar. This is the maximum potential loss in economy of design, when using the straight method compared to the other two schemes. Accounting for precompression from prestressing improves the economy of design, however.

Where the building code greatly restricts the gain in tendon stress at ULS, the application of load balancing and comprehensive methods lose their advantage. As an example, the European code[6] limits the stress gain for unbonded tendons at ULS to 100 MPa. This is a 9% gain over service condition. The allowed meager gain in stress erodes the advantage of alternative design methods, compared to the straight method.

Additional detailed information on the concept and design of post-tensioning is provided in references (Collins, M. P. et al., and Naaman, A. E., 2012).

NOTES

1. EC2 EN1002-1-1;2004(E) Section 5.10.8 (2)
2. ACI 318-19 Section 20.3.2.3.1-4.1
3. In this load combination 'total load' is considered. The coefficient of LL depends on the code case. But that of PT remains 1.00.
4. ACI 318-14 5.3.1
5. ACI 318-14 Section 20.3.2.4.1
6. EC2 EN1002-1-1:2004(E) Section 5.10.8 (2)

REFERENCES

Aalami, B. O. (2000), 'Structural Modeling of Post-Tensioned Members,' *Journal of Structural Engineering, ASCE*, V. 126, No. 2, February 2000, pp. 157–162.

Aalami, B. O. (1990), 'Load Balancing – A Comprehensive Solution to Post-Tensioning,' *ACI Structural Journal*, V. 87, No. 6, November/December, 1990, pp. 662–670.

ACI 318-19 (2019), *Building Code Requirements for Structural Concrete (ACI 318-19) and Commentary*, American Concrete Institute, Farmington Hill, MI, www.concrete.org, 623 pp.

Collins, M. P., and Mitchell, D. (1997), 'Prestressed Concrete Basics,' www.Amazon.com.

EC2 (2004), *Eurocode 2: Design of Concrete Structures – Part 1-1 General Rules and Rules for Buildings*, European Standard EN 1992–1-1:2004.

Lin, T. Y. (1963), 'Load-Balancing Method for Design and Analysis of Prestressed Concrete Structures,' *ACI Journal Proceedings*, V. 60, No. 6, June 1963, pp. 719–742.

Naaman, A. E. (2012), *Prestressed Concrete, Analysis and Design*, Techno Press 3000, Ann Arbor, MI, 1176 pp.

Chapter 4

Building codes for post-tensioning design

4.1 MAJOR BUILDING CODES

Many industrial countries have their own building codes. A number of them include the design of post-tensioned buildings. Among the international codes several are modeled after EC2[1] or ACI 318.[2] Also, BS8110 (BS8110, 1987) and TR43 (TR43, 2005) are used selectively.

The following reviews the common provisions among the major building codes, specifically EC2 and ACI 318. The common provisions are followed by specifics of each for design of post-tensioned building structures. The presentation concludes by highlighting the primary differences between the two major codes for design of post-tensioned members.

4.2 BASIC CODE CONSIDERATIONS

4.2.1 Design steps

Both codes require the design to follow three distinct steps, namely:

(i) Design for service condition: serviceability limit state (SLS).
(ii) Design for safety against collapse for specified overload: ultimate limit state (ULS).
(iii) Design for safe transfer of post-tensioning to the member at stressing (initial condition).

4.2.2 Load path

Both codes require a clear definition of structural system – load path – for transfer of load from the point of application to the supports.

For floor slabs, both codes support the breakdown of the floor into design strips in two orthogonal directions. Each design strip covers a line of supports and its tributary slab. Each design strip is detailed for the actions computed for it.

Alternatively, both codes allow for more rigorous analysis schemes.

DOI: 10.1201/9781003310297-4

53

4.2.3 Analysis schemes

A. Material response: For computation of design values, both codes recommend:
   (i) Linear elastic response of concrete.
   (ii) Member stiffness based on gross cross-sectional area.

B. Method of analysis: Both codes accept analytical schemes that follow the mechanics of solids: satisfy equilibrium and compatibility requirements. Specifically, the following is allowed:
   (i) Strip method.
   (ii) Finite element method.
   (iii) Yield line analysis.

   In addition, experimental schemes that replicate the force distribution and transfer of load in the prototype are permitted.

C. Design values: Design values are the magnitudes of the actions, such as moment, shear, axial force and torsion that are extracted from the analysis results. Design values are checked against the allowable limits of the code. They are also used to verify the adequacy of the available reinforcement, or determine the amount of reinforcement to be added.

For floor slabs, both codes extract the design values from the integration of the local values over the length of a design strip. The extraction and treatment in both codes are essentially the same, but presented somewhat differently.

For stresses, the design values are referred to as hypothetical, or computed values, since they do not refer to a specific location.

For column-supported floor slabs, ACI 318 is more specific on the extraction of the design values from the analysis results. EC2's recommendation is given in EC2's Appendix I. The outcome is essentially the same as ACI 318.

4.3 ADDITIONAL DESIGN REQUIREMENTS

4.3.1 Deflection control

Both codes specify limits on deflection for the following three conditions:

(i) Visual effects, sense of comfort.
(ii) Damage to nonstructural brittle members installed on floor.
(iii) Immediate live load deflection.

The limits set for each case by both codes are practically the same. ACI 318 limits are a multiple of 12. EC2 limits are a multiple of 10.

EC2 permits the compliance with deflection limit be waived, if in the judgment of the engineer the estimated value is not critical for the condition at hand.

4.3.2 Crack control

Crack control is handled differently for one-way and two-way systems.

(i) For beams and one-way slabs, both codes allow cracking in service condition. Both codes offer provisions to account for the impact of cracking on member deflection.
(ii) For column-supported (two-way) floor systems, EC2 permits cracking for service condition. The crack width permitted in design depends on the exposure of the member and whether the member is reinforced with bonded or unbonded tendons. Computed probable crack width can be controlled by addition of rebar.
(iii) ACI does not permit computed cracking for service condition. Likelihood of cracking is controlled through setting limits on the hypothetical tension stresses of the design strips.

4.3.3 Member ductility

Both codes limit the amount of tension reinforcement of a section in flexure. This ensures that when overloaded, the section failure is initiated by yielding of its reinforcement. The reinforcement limits set in the two codes are close to one another.

4.4 EC2 SPECIFIC CODE PROVISIONS

4.4.1 Minimum reinforcement

EC2 specifies the minimum amount of reinforcement based on the cross-sectional geometry of the member. The minimum can be a combination of prestressed and non-prestressed reinforcement.

4.4.2 Maximum reinforcement

EC2 sets a limit on the maximum amount of combined non-prestressed and prestressed reinforcement in member section.

4.5 ACI 318 SPECIFIC CODE PROVISIONS

4.5.1 Minimum precompression

For column-supported floor slabs ACI sets a minimum value for the average precompression from prestressing. This translates to a minimum provision of reinforcement in column-supported slabs. EC2 achieves the same objective by specifying a minimum cross-sectional area of reinforcement, as opposed to precompression.

4.5.2 Arrangement of prestressing tendons

For column-supported floors, in one direction ACI limits the maximum spacing between adjacent tendons.

EC2 offers neither provision, nor limitation, for arrangement of post-tensioned tendons.

4.5.3 Transfer of column moment to slab

Where at slab-column joint the slab is required to resist all or part of the column moment at slab-column connection, ACI has a specific recommendation for detailing the slab in the vicinity of the slab-column location. It is referred to as 'transfer of unbalanced moment.'

4.5.4 Minimum bar length

ACI 318 specifies the minimum length and arrangement of non-prestressed reinforcement required for service condition. There is also provision for bar lengths if the bars are required to contribute to the safety of the member.

4.6 NOTABLE DIFFERENCES BETWEEN EC2 AND ACI 318

4.6.1 Strength and material factors

Both ACI and EC2 recognize the approximate nature of the computational methods in predicting the strength of an over-loaded member.

EC2's provision is based on the accuracy with which the mechanical properties of the material used in a structural member can be predicted, irrespective of whether the member is in bending or axial force. For this reason, different reliability 'material factors' are used for different materials.

As an example, the material factors for concrete and prestressing steel are:

For concrete:

$$\gamma_c = 1.5 \tag{4.1}$$

For prestressing and non-prestressed reinforcement:

$$\gamma_s = 1.15 \tag{4.2}$$

The characteristic properties of the material, such as 28-day concrete strength, are divided by 1.5 in the computation of section strength. But, for non-prestressed reinforcement the measured strength is divided by 1.15 for greater reliability in predicting the property of reinforcement.

ACI 318's provisions are based on the presumption that the central cause of the approximation is the accuracy in the computational method used to predict the capacity of a member under increased loading.

As an example, design engineers can predict relatively well the strength of a beam under bending. But if the same member is loaded axially, the calculated axial force for the member failure is not as close. For this reason, the code provides the concept of 'strength design factor.' It is a coefficient by which the calculated strength values are adjusted to better predict the member's response based on the design response of the member to resist the applied load.

As an example, the strength reduction factor φ for bending and axial force is:

For bending:

$$\varphi_{bending} = 0.9 \tag{4.3}$$

For axial:

$$\varphi_{axial} = 0.7 \tag{4.4}$$

4.6.2 Contribution of prestressing to member strength

Both codes recognize that when overloaded, prestressing tendons stretch beyond their service condition. The stretching provides greater moment capacity to resist the demand. Both codes allow for increase in tendon stress at ULS.

For bonded tendons, allowance for stress increase at ultimate in EC2 is almost two times the limit given in ACI 318. This provision favors the application of bonded tendons using EC2. For unbonded tendons the EC2 allowance for stress increase is practically negligible.

The ACI 318 allowance for stress increase for unbonded tendons can be over four times that allowed in EC2. This significantly favors the application of unbonded tendons.

4.6.3 Punching shear

There is little correlation between the punching shear provisions of EC2 and ACI 318. Several notable differences are as follows:

A. Tension bars over the column: EC2's punching shear collapse mechanism is based on the strut-and-tie concept, where tension bars over the column are an essential element of the strength mechanism. Without tension reinforcement over the column there is no punching shear capacity.

ACI-318's capacity is based on value of fictitious stresses over the face of an assumed fictitious collapse surface. It does not explicitly include the contribution of tension bars over the support.
B. EC2 assumes the first critical section to be at the face of the column; the next section $2d$ away from the face of the column, where d is the distance from extreme compression fiber to center of tension. ACI-318 assumes the first critical section to be $0.5d$ from the face of the column.
C. ACI-318 considers the strength in punching shear to be subject to the 'magnitude' of the available moment in 'one-direction.' The magnitude of column moment in one direction only enters the computation of the hypothetical design stress at the face of the assumed collapse mechanism. Biaxial moment effects are not accounted for. Assumed stresses from moments in each principal direction are computed and handled separately.

EC2 allows for the 'presence' of column moment at column-slab connection, but not the magnitude of the moment. The presence – not the value – of moment reduces the allowable punching shear value.
D. The assumed geometry of the critical section for punching shear strength check is widely different between the two codes.
E. ACI 318 requires a fraction of the column-slab moment to increase to punching shear demand. EC2's punching shear check does not depend on the magnitude of the column-slab moment. EC2 allows for the presence of moment – not its magnitude.
F. The computed punching shear design capacity between the two codes can be widely different (Widianto, Oguzhan Bayrak, 2009).

Neither ACI nor EC2 strictly follow the mechanics of solids in predicting the punching shear capacity of a column-slab location. They rely primarily on test results.

4.6.4 Application of cracking moment

Both codes require the moment capacity of a section exceed the moment that can initiate its cracking. But the application of the requirement is different. The focus of ACI is its application to members that are reinforced with bonded tendons; EC2 on the other hand specifies the requirement for unbonded tendons.

4.7 EUROPEAN CODE FOR DESIGN OF POST-TENSIONED MEMBERS

The current European code for design of concrete structures is EN 1992-1-1 Part 1-1, updated 2004. In the following, for brevity it is referred to as EC2. In the footnotes it is quoted as EC2 EN 1992-1-1:2004(E).

Building codes for post-tensioning design 59

EC2 covers the design of both conventionally reinforced and pre- or post-tensioned members of building structures.

What follows is limited to provisions that directly impact the design of post-tensioned members in common building construction. For details of the provisions consult the full text of EC2.

The provisions extracted are organized in the sequence that a design engineer is likely to refer to in the course of design of a post-tensioned building.

4.7.1 EC2 code compliance basics

For design of post-tensioned members, the focus of the serviceability check of the European code is crack control, followed by deflection. The level of compliance of each depends primarily on its perceived consequence as opposed to prescriptive stipulations.

Measures for crack control are triggered by the magnitude of a hypothetical (representative) extreme fiber tensile stress for given load combinations. Once the threshold is exceeded, check for cracking is initiated. There are recommended measures to control cracking.

The estimate of deflection for practical and common conditions is based on defined load combinations and gross cross-sectional properties. Allowance for creep and cracking is required, but for limited conditions.

The code is detailed, requiring diligence to navigate the recommendations for specific application. The emphasis of what follows is on the provisions that are applicable to common residential and commercial post-tensioned buildings.

First, the parameters (thresholds) that govern the design are reviewed. Next, by way of flow charts, the steps for the serviceability compliance of post-tensioned floors are detailed. The serviceability check is followed by design for safety. The design is concluded by checking the safe transfer of prestressing at stressing.

4.7.2 Stress threshold

For service condition EC2 sets limits for the extreme fiber stress of concrete; stress in non-prestressed reinforcement; and stress in prestressing steel. For practical building structures, only the extreme fiber stress in concrete applies.

Gross cross-sectional properties and linear elastic material properties are used to calculate the demand forces.

For beams and one-way slabs, simple beam theory is used to determine the extreme fiber stress. The hypothetical (representative) stress for both the slab and the beam is calculated using the beam theory given by:

$$f = \frac{Mc}{I} + \frac{P}{A} \qquad (4.5)$$

where

M = demand moment from code specified load combination;
I = second moment of area of the section resisting M;
c = distance of the farthest fiber from the section's centroid;
P = axial force on the section; and
A = cross-sectional area of the section.

For column-supported floors, EC2 recognizes that the distribution of stress across the slab is nonuniform. Stresses are higher near the column support and fall with distance from the column.

For each panel a representative hypothetical stress is used to assess the probability of crack formation and the requirement for crack control.

A threshold is defined for the hypothetical stress of each panel. Once the hypothetical stress exceeds the threshold, other provisions are triggered. The triggered provisions provide two options. Either estimate the width of the probable cracks for its compliance with the allowable, or use prescriptive measures to control the computed width of the cracks.

The determination of the hypothetical stress is explained next.

4.7.2.1 Hypothetical extreme fiber stress

The definition and evaluation of the hypothetical stress for the design of column-supported slabs is given in Appendix I[3] of the code. The following excerpts from Appendix I explain the procedure.

- Analysis may be based on equivalent[4] frame, grillage, finite elements, or yield line.
- Stiffness is based on gross cross section.
- For vertical loads, slab stiffness is based on full width of the column tributary.
- 100% of load is to be used for analysis in each direction.

EC2 recommends subdividing the floor into design strips in two orthogonal directions. Each design strip includes a line of supports and the associated slab tributary. The concept is explained in greater detail in Chapter 1.

For each design strip, design sections across the entire width of the strip are selected.

For each design section, the total value of forces acting on that strip is calculated. The total value of the moment and the axial force on the gross cross-sectional face of the design section is used to arrive at a single representative extreme fiber stress. This single stress is referred to as hypothetical, since it does not apply to a specific point.

For code compliance, EC2 recommends a larger value for the hypothetical stress than the 'average' extreme fiber stress obtained from direct application of the total design strip force to the total cross-sectional geometry of the design strip.

The enlarged value of the hypothetical stress recommended in EC2 is based on the 'column strip/middle strip' concept for distribution of design strip values.

Figure 4.1 from Appendix I of EC2 shows the recommended subdivision of the panel into column strip and middle strips.[5]

EC2 leaves it to the judgment of the designer to assign from 60% to 80% of the entire design strip moment to the column strip. The balance is distributed equally between the middle strips.

EC2 provides Table 4.1 as a guide to the design engineer for breakdown of the design values of the design strip between the column strip and middle strips.

For conventionally reinforced concrete, ACI 318[6] recommends 65%/35% ratio for assignment of moment between the column strip and middle strips. In the absence of specific recommendations, the default values for the preceding EC2 table may be based on 65%/35% distribution, keeping the option to modify the ratio, if required.

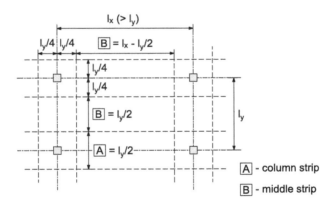

Figure 4.1 Division of a flat slab panel into column and middle strips.

Table 4.1 EC2 recommended values for breakdown of design strip values between column strip and middle strips

	Negative moment	Positive moment
Column strip	60–80%	50–70%
Middle strip	40–20%	50–30%

Note: Total negative and positive moment to be resisted by the column and middle strip together add up to 100%.

Source: Reproduced from Table I.1 of EC2[a].

[a] Eurocode 1, Part 2.1 (ENV 1991-2-1); Table I-1

The ACI 318 recommended breakdown for conventionally reinforced concrete slabs is within the range of parameters of the preceding table.[7]

Figure 4.2 shows a typical computed distribution of tributary moment at the face of a column-supported floor together with the idealization recommended by EC2.

The following determines to what extent the EC2 suggested breakdown of the tributary moment increases the otherwise average stress – the average being the total moment applied to the entire section, similar to ACI 318. The objective is to simplify the application of EC2 provision to practical design, while keeping strict compliance with the code. The simplification does away with the column strip/middle strip by applying the total strip value to total strip section and increasing the outcome by a given coefficient.

Let k_c be the fraction of the tributary moment (M_t) allocated to the column strip.

Average value of moment for the column strip for hypothetical stress compliance is:

$$\text{Average stress} = (k_c\, M_t)/A_c$$

where A_c is the cross-sectional area of the column strip. It is generally one half the area of the total strip.

The average moment of the design strip is (M_t/A_t).

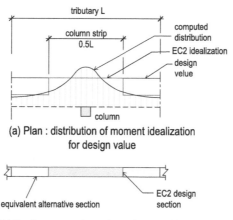

Figure 4.2 Idealization of tributary moment for EC2-based designs.

Building codes for post-tensioning design 63

From the foregoing, the increase in average bending stress resulting from the consideration of the entire tributary for definition of the hypothetical strip is:

Increase in bending stress = $[(k_c\, M_t)/A_c\,]/[\,(M_t/A_t)] = k_c\,(A_t/A_c)$

Assuming $k_c = 0.65$ which conforms[8] with the EC2 Table 4.1, and A_c one half the total tributary area A_t, the increase in average stress is:

Increase in average stress = $k_c\,(A_t/A_c) = 0.65 \times 1/0.5 = 1.30$

The preceding leads to the conclusion that if the entire tributary and the entire moment are used to calculate the hypothetical stress, the calculated value must be multiplied by 1.30 to conform with the EC2 requirememnt. This conclusion greatly simplifies the designs based on EC2, while maintaining code compliance. More importantly, it improves the accuracy of the computed value for the reason described next.

For post-tensioned members, the hypothetical stress used for design consists of the sum of bending and axial stresses:

$f_{hpothetical} = f_{bending} + f_{axial}$

The axial contribution of stress to the hypothetical stress is not subject to the column strip/middle strip distribution. In the common case, the axial stress is distributed somewhat uniformly over the entire tributary. For this reason, for post-tensioned floors ACI 318 does not support the column strip/middle strip concept.

In summary, for compliance with EC2 recommendations in calculating the hypothetical tensile stresses, the following simplified approach applies.

- Base the design on forces calculated for the entire tributary of design strips.
- Apply the entire tributary moment and the entire tributary axial force to the entire cross-sectional geometry of the design strip to calculate the hypothetical extreme fiber stresses.
- Multiply the calculated extreme fiber stress by 1.30 to arrive at the EC2 required hypothetical stress threshold for probability of crack formation.

4.7.2.2 Design crack width

EC2 is flexible regarding the allowable crack width and its control.[9] For each design, depending on the anticipated exposure of the member to the elements of corrosion, a recommended design crack width (w_k) applies.

The selection of 'design crack width w_k' is required if the design permits crack formation; otherwise not.

The following excerpts from the code[10] explain:

(1) Cracking shall be limited to an extent that will not impair the proper functioning or durability of the structure or cause its appearance to be unacceptable.
(2) Cracking is normal in reinforced concrete structure subject to bending, shear, torsion or tension resulting from either direct loading or restraint or imposed deformation.
(3) Cracks may also arise from other causes such as plastic shrinkage or expansive chemical reactions within the hardened concrete. Such cracks may be unacceptably large but their avoidance and control lies outside the scope of design described in this book.
(4) Cracks may be permitted to form without any attempt to control their width, provided they do not impair the functioning of the structure.
(5) A limiting calculated crack width, w_{max}, taking into account the proposed function and nature of the structure and the cost of limiting cracking should be established.
(6) The recommended values for maximum crack width w_{max} for different environmental exposures are given in EC2.[11]

For common residential and commercial building projects, the recommendation of EC2 is summarized in Table 4.2.

Table 4.2 is reproduced for completeness. The recommendations of the table cannot be applied strictly to common building construction, as explained in the following.

The evaluation of crack width for code compliance is based on (i) computed extreme fiber stress at the 'point,' where the crack is likely to initiate; and (ii) the stress in the reinforcement close to it. The position of reinforcement in-place necessary to arrive at the probable crack width is not always reliably known at design stage. Figure 4.3 is an example of a post-tensioned column-supported floor ready to receive concrete. The details of the reinforcement in place – as evident from the figure – are not predictable at design stage with the accuracy required for the computation of probable crack width (Figure 4.4).

Table 4.2 Design crack width recommendation[a] (w_k)

	Unbonded system Quasi-permanent	Bonded system Frequent
Common building frames not exposed to weather and moisture (building interior)	0.4 mm	0.2 mm
Common building frames exposed to weather and occasional wetting	0.3 mm	

[a] Eurocode 1, Part 2.1 (ENV 1991-2-1) EC2 Table 7.1N.

Figure 4.3 View of a column-supported post-tensioned floor reinforced with bonded tendons ready to receive concrete.

For post-tensioned floors, it is recommended to follow the code-stipulated stress threshold in controlling the formation of probable cracks, as opposed to calculating the crack width. EC2 provides detailing options for crack control as an alternative to crack width calculation. The detailing options are referenced in the flow charts of this chapter.

4.7.3 Load combinations

The load combinations for gravity design include dead, live and effects of post-tensioning.

The fraction of design live load in the load combination depends on the occupancy of the floor. Depending on the occupancy, EC2 recommends inclusion of the fraction of design live load given in Table 4.3. For common residential and commercial floors 30% of the design live load is recommended for the quasi-permanent (sustained) load combination.

The following three load combinations commonly apply.

Characteristic (total)

$$U = 1.0DL + 1.0LL + 1.0PT \tag{4.6}$$

Quasi permanent (sustained)

$$U = 1.0DL + 0.3LL + 1.0PT \tag{4.7}$$

Frequent

$$U = 1.0DL + 0.5LL + 1.0PT \tag{4.8}$$

where PT represents forces in the member resulting from post-tensioning. PT includes member forces from the reactions generated at the support of the member from post-tensioning, if any.

66 Post-tensioning in building construction

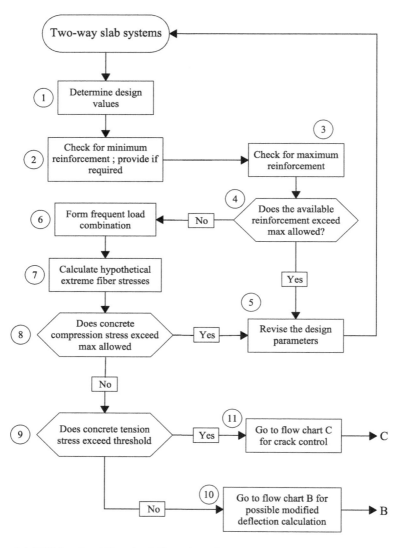

Figure 4.4 EC2 Serviceability check flow chart A.

Table 4.3 Fraction of design live load to be considered as 'sustained; quasi-permanent.'[a]

Occupancy	Fraction of design live load ψ
Dwellings and offices	0.3
Shopping; congested areas	0.6
Storage	0.8
Parking	0.6

[a] Eurocode 1, Part 2.1 (ENV 1991-2-1)

The requirements for the characteristic load combination do not generally apply to common building structures. The other two combinations are commonly used.

4.7.4 Serviceability design thresholds; design limits

The design thresholds (limiting values) of EC2 are expressed in terms of the properties of the material used. These are:

A. **Tensile strength**: Cracking from extreme fiber tensile stress in concrete[12]

$$f_{ctm} = 0.3 f_{ck}^{(2/3)} \text{ for concrete} <= C50/60$$

[13]

f_{ck} = compressive strength of concrete cylinder at 28 days

$$f_{ctm} = 2.12 \ln[1+(f_{cm}/10)] \text{ for concrete } > C50/60$$

$$f_{cm} = f_{ck} + 8$$

C50/60 refers to concrete's characteristic cylinder to cube strength in MPa

where f_{cm} is the mean compressive strength of concrete at 28 days in MPa

B. **Compressive strength**: Threshold for extreme fiber hypothetical compressive stress (f) in concrete[14]

$$f = 0.6 f_{ck}$$

C. **Allowable tensile stress (f) in prestressing steel**[15]

$$f = 0.75 f_{yk}$$

Allowable tensile stress in non-prestressed reinforcement

$$f = 0.80 f_{yk}$$

Other thresholds for serviceability design of post-tensioned members in building construction are crack width and deflection.

Acceptance of deflection and/or cracking is subject to the consequence of each on the extended performance of the member. The allowable limit varies from case to case. It is subject to evaluation and judgment of the design engineer and the owner.

4.7.5 Serviceability design flow charts

EC2 covers pre- and post-tensioned members for a wide range of applications. Design of post-tensioned floor slabs and beams used in building construction are not the focus of the code.

A newcomer to EC2, planning to design a building structure, needs to navigate through the seemingly labyrinth provisions of the code to identify the items that are: (i) relevant to the design at hand; and (ii) are likely to govern the design.

The following aid is the excerpt of the code provisions applicable to design of post-tensioned building structures – specifically slabs. It is organized to satisfy the serviceability of the code (SLS). Each item is referenced to the respective provision of the code.

The design example provided in Chapter 5 illustrates the application.

Start with flow chart A and conclude with flow chart B or C in Figures 4.5 and 4.6, depending on the parameters of design.

The following explains each of the design steps identified in the preceding flow charts:

(1) **Determine design values**

At this stage, it is assumed that the analysis of the floor, including its post-tensioning, is complete; design strips and design sections have been identified; the cross-sectional geometry of each design section is known; and the design forces for each section are calculated. The design is ready for code compliance check.

The design actions generally include moments, shears and axial forces. The serviceability check reviews the status of each design section in regard to the serviceability requirements of the code.

(2) **Minimum overall reinforcement**

The objective of minimum reinforcement is crack control from shrinkage and temperature effects. The crack control from applied loads is covered in other parts of the flow chart.

Each design section shall be checked for the total amount of its prestressed and non-prestressed reinforcement. The total amount of reinforcement in each design section shall neither be less nor more than the prescribed value of the code.

At this stage of design, the layout and amount of prestressing at each design section is known. Also, the designer may have pre-specified non-prestressed base reinforcement, such as top or bottom mesh bars at specific locations. If the total amount of provided reinforcement is less than the minimum required, the shortfall should be added.

The minimum amount of reinforcement depends on two considerations:

(i) The cross-sectional geometry of the design section.
(ii) For beams reinforced with unbonded tendons, the ratio of the beam's moment capacity to its cracking moment.

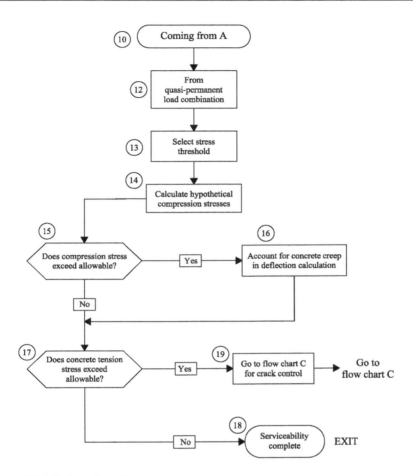

Figure 4.5 EC2 flow chart B.

(a) Based on cross-sectional area: The minimum reinforcement A_{smin} is given in EC2 Section 9.2.1.1 for beams, and Section 9.3.1.1 for slabs. Both sections use the following relationship.

$$A_{smin} \geq \frac{0.26 b_t d f_{ctm}}{f_{yk}} \geq 0.0013 b_t d \tag{4.9}$$

where
d = depth from compression fiber to the centroid of non-prestressed steel. The distance (d) refers to where non-prestressed steel is either located or will be positioned, where needed;
b_t = mean width of the tension zone;
f_{ctm} = axial tensile strength of concrete; and

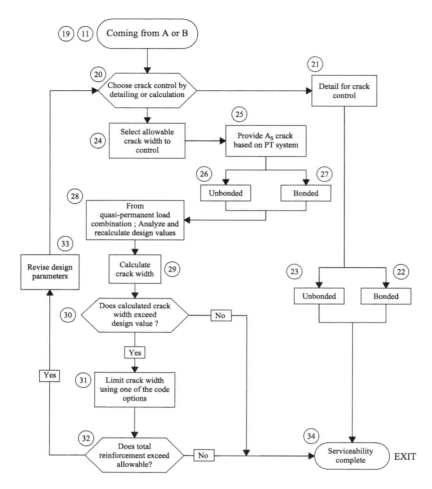

Figure 4.6 EC2 flow chart C.

f_{yk} = yield stress of non-prestressed reinforcement; f_{pk} is used in lieu of f_{yk}, where section is prestressed.

If both prestressed and non-prestressed steel are present in the same design section, the weighted average of their characteristic strengths is used. The intent of this provision is crack control arising from shrinkage and temperature changes.

The provided reinforcement to be compared with the minimum required is given by:

$$A_{sprov} = A_s + A_{ps} \times \frac{f_{pk}}{f_{yk}} \qquad (4.10)$$

The reason the area of prestressed steel (A_{ps}) is enhanced by the ratio given above is to recognize the precompression it provides in addition to the restraint to shrinkage and cracking strains. Precompression in the right amount, on its own, can be adequate to mitigate shrinkage and temperature cracking. Hence, its presence is accounted for in the preceding equation. If the provided (A_{sprov}) reinforcement is less than the minimum determined (A_{smin}), increase the reinforcement to the minimum value.

(b) Based on moment capacity: The amount of prestressed and non-prestressed reinforcement at a section of a beam reinforced with unbonded tendons shall be adequate to develop a moment capacity at that section not less than 1.15 times the cracking moment computed for the same section.[16]

This check is deferred and handled at the end of the safety check.

(3) **Maximum overall reinforcement**

Each design section is checked for not exceeding the maximum total reinforcement (A_{smax}), using EC2 Section 9.2.1.1 for beams and Section 9.3.1.1 for slabs. Both sections use the following relationship.

$$A_{smax} = 0.04 A_c \tag{4.11}$$

where A_c is the gross cross-sectional area of the design section.

If both prestressed and non-prestressed steel are present, the weighted average of their characteristic strengths is used.

$$A_{sprov} = A_s + A_{ps} \times \frac{f_{pk}}{f_{yk}} \tag{4.12}$$

(4) (5) **Revise design parameters**

If the provided steel (A_{sprov}) is more than the maximum allowable A_{smax}, the design must be revised.

(6) **Form frequent load combination**[17]

Serviceability check is carried out for both the frequent and quasi-permanent load combinations. Start with frequent load combination by selecting the appropriate value of ψ, based on the occupancy of the floor.

For residential and office buildings, the load combination is:

$$U = 1.00 Selfweight + 1.00 DL + 0.50 LL + 1.00 PT \tag{4.13}$$

(7) **Calculate the hypothetical extreme fiber stress**

In column-supported slabs, the local concrete stress varies significantly from point to point. Stresses are high over the supports and drop rapidly with distance away from it. In practice, rather than focusing on or calculating stresses at points along a design strip, a representative (hypothetical) stress is calculated for each design section. The hypothetical stress is used as an indicator for crack control. The computation of hypothetical stress is outlined in earlier sections of this chapter.

(8) **Compare hypothetical compression stress with threshold value**

In practice, it is rare but feasible that the hypothetical compressive stress in concrete exceeds the set threshold. In this case, the parameters of design have to be revised. The design must be re-done from the beginning.

Strictly speaking, at this stage stresses in prestressing steel and non-prestressed steel must also be checked. If computed values exceed the selected thresholds for design, the design must be revised. The condition does not govern the common residential and commercial buildings. It is included herein for completeness. For common conditions, this check is skipped.

(9) **Does hypothetical concrete tension stress exceed the threshold limit?**

The outcome of comparison determines whether additional checks for crack width must be followed, or whether deflection calculations should consider concrete creep.

Depending on the outcome, the serviceability check follows flow chart B or C.

(10) **Follow flow chart B for possible modified deflection calculation**

If the hypothetical stresses do not exceed the design threshold, the serviceability check will be continued following flow chart B. Flow chart B covers the 'quasi permanent' load combination. If concrete compressive stresses are high, it requires to allow for nonlinear concrete creep in the calculation of deflection.

(11) **Follow flow chart C for crack control**

If the hypothetical tensile stresses exceed the design threshold, probability of undesired crack formation exists. Non-prestressed bonded reinforcement may have to be added; or probable crack width must be calculated and checked to be less than the design value (w_k).

FLOW CHART B

(12) **Form quasi-permanent (sustained) load combination**[18]

The load combination for common building construction is:

$$U^{19} = 1.00DL + 0.30LL + 1.00PT \qquad (4.14)$$

(13) **Select concrete stress thresholds**

Concrete stress thresholds for 'quasi-permanent' load combination are:

Maximum threshold for hypothetical compressive stress is $(0.45f_{ck})$.[20]
Maximum threshold for hypothetical tensile stress is $(f_{ct,eff})$.[21]

(14) **Calculate the hypothetical extreme fiber concrete stresses**

The calculated hypothetical extreme fiber stresses are compared with the threshold values.

(15) **Compare the hypothetical compressive stress with threshold value**

If the hypothetical compressive stress exceeds the threshold, allowance should be made for creep of concrete in calculating the deflection of the member.

(16) **Account for concrete creep deflection calculation**

This scenario is triggered, if under quasi-permanent load combination compressive concrete stress exceeds the set threshold. Deflection calculation must allow for increased concrete creep.

The code recommends using a non-linear procedure for the creep component of deformation in deflection calculation.

(17) **Does the hypothetical tensile stress exceed the threshold?**

The hypothetical extreme fiber tensile stresses under the quasi-permanent load combination are checked against the threshold to determine the likelihood and nature of cracking.

(18) **Stresses do not exceed the threshold**

If the hypothetical extreme fiber tensile stress does not exceed the threshold, the design is deemed to satisfy the serviceability requirements of the code.

(19) **Assessment of cracking and its control become necessary**

Higher value of hypothetical tensile stresses calculated point to the likelihood of probable cracking and the requirement for its control.

Go to flow chart C for crack control.

FLOW CHART C

(20) **Choose method of crack control**

For crack control EC2 offers two options:

Either check the available reinforcement and add where it is not adequate for code-required crack control; or calculate the probable crack width, and demonstrate that it does not exceed the design value (w_k). If the calculated crack width exceeds the design value, add reinforcement to reduce it to the design value.

The crack width is highly dependent on (i) local stress, and (ii) the availability and distribution of bonded reinforcement in the vicinity of crack initiation. Neither of the two conditions are readily applicable in common building floor construction. Specifically, the hypothetical extreme fiber tension stress used in design does not correlate with the point of probable crack formation. The in-place position of next bonded reinforcement to the probable location of cracking is also not reliably known. It is recommended to use the added rebar option.

(21) **Detail for crack control**

In this option, crack width calculation is not required. Instead, the member is checked for the amount and distribution of bonded reinforcement. For this option, bonded reinforcement shall not be less than the amount specified by code.

The amount of minimum bonded reinforcement for crack control is given in Tables 7.2N and 7.3N of EC2. The entry value to these tables requires the magnitude of service condition stress in the non-prestressed reinforcement at the location where cracking is likely to occur.

Again, this information is neither available at this stage in design, nor can it reliably be obtained for common column-supported slab systems. The alternative is to use the recommendation of TR43 (TR43, 2005), which in effect is the translation of the EC2-intended provision to practical design of post-tensioned floor systems.

The following is the TR43 recommendation for crack control, where the hypothetical extreme fiber tension stress in span exceeds the threshold.

The requirement of crack control minimum reinforcement is met by providing bonded reinforcement in the tensile zone of the concrete section. The required bonded reinforcement shall not be less than A_{scrack}.

$$A_{scrack} = \frac{N_c}{(5/8)f_{yk}} \quad (4.15)$$

where

N_c = tensile force over the tension zone of the concrete section of the design strip (Figure 4.7) and

f_{yk} = yield stress of the reinforcement used for crack control.

Figure 4.7 shows the distribution of stress through the concrete section and the tensile force in the tension zone of the section. In this figure h is the depth of the member and c the depth of compression zone.

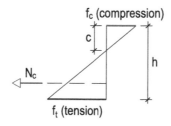

Figure 4.7 Computed distribution of stress over concrete section.

(22) **Bonded system crack control reinforcement**
Where the bonded system is used, the cross-sectional area of the bonded reinforcement is accounted for.

$$A_{scrack} = A_s + A_{ps,bonded} \qquad (4.16)$$

In this case, the contribution of bonded reinforcement ($A_{s,bonded}$) will not be enhanced by its higher yield stress to the non-prestressed reinforcement (A_s). The contribution of precompression to crack control has already been accounted for in the computation of hypothetical tensile stresses.

(23) **Unbonded system crack control reinforcement**
Where the unbonded system is used, the entire area of the required A_{scrack} must be provided by non-prestressed reinforcement.
A_{scrack} is given by Equation (4.15) of EC2.

(24) **Select allowable crack width**
The allowable crack width for each floor depends on the anticipated exposure of that floor to corrosive elements, or the visual impact of probable cracks.

The exposure classifications[22] and the recommended values are given in Table 7.1N of the EC2 code. The values range from 0.2 to 0.4 mm.

For members reinforced with unbonded tendons, the most common selection is 0.3 mm. For members reinforced with bonded systems, the suggested value for most exposures is 0.2 mm.

(25) **Provide/check minimum cracking reinforcement**
Depending on the post-tensioning system used, EC2 recommends minimum bonded reinforcement.

(26) **Minimum crack control reinforcement for unbonded systems** (A_{scrack})
Since the hypothetical tension stress at the farthest fiber exceeded the lower allowable threshold, control of potential cracks becomes necessary for the affected design sections.

The requirement of minimum reinforcement for crack control is met by providing bonded reinforcement in the tensile zone of the concrete section. The minimum amount of the bonded reinforcement necessary under this provision will be increased if crack width controls that are in the other steps of the flow chart conclude with additional reinforcement.

The requirement will be expressed herein in its simplified format applicable to floor systems. EC2 provides detailed instruction for crack width calculation and the required reinforcement.

$$A_{scrack} = \frac{N_c}{f_{yk}} \qquad (4.17)$$

where

N_c = tensile force over the tension zone of the concrete section (Figure 4.7), and

f_{yk} = yield stress of the reinforcement used for crack control.

Note that the denominator is much higher than the value suggested for a similar condition in either ACI 318 or TR43. Detailed calculation for probability of crack formation and the required reinforcement is given in EC2.[23]

(27) **Minimum crack control reinforcement for bonded systems** (A_{scrack})

For bonded systems, the required minimum area of reinforcement is the same as described for the unbonded systems, with the difference that the available area of bonded tendons will be included in the computations.

EC2 specifies maximum 300 mm[24] spacing for crack control reinforcement. The minimum area of reinforcement for crack control A_{scrack} is given in EC2 EN 1992-1-1:2004(E), Section 7.3.2(3).

(28) **Calculate the design values for quasi-permanent (sustained) load combination**

The probable crack width is estimated using the quasi-permanent load combinations. Calculate the design values.

(29) **Calculate probable crack width**

Using the procedure detailed in EC2,[25] and the contribution of non-prestressed reinforcement from the previous steps, the serviceability check continues with the computation of the probable crack width (w_k) for each design section. The calculation of the probable crack width (w_k) is outlined in EC2 EN 1992-1-1:2004(E), Section 7.3.4.

(30) **Does the calculated crack width exceed the design value (w_k)?**

If it does not, exit. The serviceability check is satisfactorily completed.

If the calculated crack width exceeds the allowable, limit the crack width by adding reinforcement detailed in the following step.

(31) **Add reinforcement to control crack width**

If the computed probable crack width from the previous step exceeds the value selected for design, EC2 provides two remedial options.

Either add reinforcement using Section 7.3.4 relationship 7.9 of EC2 to limit the probable crack width (w_k); or, select non-prestressed bar diameter and spacing according to EC2 Table 7.2N or Table 7.3N.

(32) **Does the total amount of reinforcement exceed the allowable?**

At this stage, having added reinforcement for crack control, in theory, the total amount of reinforcement can exceed the maximum allowable, hence triggering a repeat check for maximum reinforcement.

The check for maximum reinforcement is outlined in step 3 of the flow chart.

This is a rare condition. It is noted here for completeness.
(33) **Revise design parameters**
If the total amount of reinforcement exceeds the allowable value determined in flow chart A, the parameters of design, such as member thickness and concrete strength must be revised and the design repeated.
(34) **Serviceability requirements of the code are met**
Exit the serviceability check.

4.7.6 Strength design

For post-tensioned floor slabs, at this stage of design the amount and location of post-tensioning are known.

It is also likely that non-prestressed reinforcement has been added to meet the serviceability requirements of the code.

Further, the local practice may have included adding bottom or top mesh reinforcement, in addition to that required by computation.

In summary, unlike conventionally reinforced floors, post-tensioned floors possess a base strength capacity that often exceeds the design moment required for the safety check of the code. For this reason, it is expeditious to start with the determination of the base capacity of the member. Add reinforcement, where base capacity falls short of demand moment.

4.7.6.1 Load combination

The strength load combination for common residential and commercial post-tensioned members is:

$$U = 1.35DL + 1.60LL + 1.00HYP \tag{4.18}$$

where HYP is the force effects resulting from the reactions of post-tensioning on the supports of the member – hyperstatic actions.

Alternative load combinations, where the HYP component is replaced by the post-tensioning forces, are given in Chapter 3.

4.7.6.2 Cracking moment and flexural strength

To avoid failure of a member's section at the initiation of cracking, the reinforcement in the section must develop moment capacity for the section not less than the cracking moment M_{cr} of the same section.

EC2[26] specifies a safety factor of 1.15 for bending strength of beams reinforced with unbonded tendons at initiation of cracking.

Details for the computation of the cracking moment and the required reinforcement is given in reference (Aalami, B. O., 2021).

4.7.6.3 Punching hear

The punching shear formulation of EC2 along with a numerical example are given in reference (Aalami, B. O., 2014).

4.7.6.4 Detailing

EC2 does not provide recommendations specific to construction detailing of post-tensioned floors.

4.8 ACI 318 PROVISIONS FOR POST-TENSIONED FLOORS

The following presents the ACI 318 provisions of the code that engineers are likely to encounter in the design of post-tensioned floors. The code sections are listed in the sequence of a typical design. For full coverage refer to the text of the code.

4.8.1 Floor slab categorization and geometry

ACI 318 breaks the design of floor slabs into the following categories.

- A. **One-way slabs**: These are defined and covered in Chapter 6 of ACI-318. They are reviewed by way of an example in Chapter 6 of this book.
- B. **Two-way slabs**: Two-way slabs refer primarily to column-supported floors. The design of column-supported floors is covered in Chapter 8 of ACI-318. A detailed design example is given in Chapter 5 of this book.
- C. **Beams**: Beams are handled in Chapter 9 of ACI 318. Chapter 6 of this book is a detailed example of a beam frame based on ACI 318.

4.8.2 Define loads

Floor slabs are commonly designed for gravity loads and checked for other conditions, such as wind and earthquake. Where applicable, slabs are also checked for temperature and vibration.

ACI 318 does not specify the magnitude of the design loads, nor the reduction of live loads. The current version of ASCE-7[27] should be consulted.

4.8.3 Validate sizing and material properties

Before engaging in detailed numerical analysis, it is important to make sure that the slab thickness, concrete strength and the arrangement and size of the supports are adequate for the specified load.

4.8.3.1 Punching shear check

This initial check is intended to identify the column supports that cannot be designed to meet the required punching shear capacity. A second check toward the end of the design concludes the punching shear design.

Estimate the punching shear force of the critical column supports, using column tributaries. Assume each column supports the weight of the slab region halfway between itself and the next support around it.

Use the following load combination to estimate the initial punching shear force demand:

$$V_u = 1.20DL + 1.60LL \qquad (4.19)$$

Strictly, the preceding relationship must also include the hyperstatic contribution of post-tensioning. But, at this stage of design the value is not known. Also, the inclusion of post-tensioning in the general case does not significantly change the design support reaction for most columns.

A. **Check for maximum stress:** Verify that the geometry of the support does not lead to punching shear stress that exceeds the limit permitted in ACI 318.

   Check the average punching shear demand v_u at the first critical section.

$$v_u = V_u / A_c \qquad (4.20)$$

   where
   v_u = punching stress demand; and
   A_c = surface area of the first critical section.
   Compare the average punching shear stress demand v_u with the maximum allowable value. For post-tensioned floors the demand shall not exceed the following:[28]

$$v_u \le 0.67\varphi\sqrt{f_c'} \qquad (4.21)$$

   For gravity and wind design[29] $\varphi = 0.75$.

B. **Check for punching shear reinforcement:**
   If the objective of design is to avoid installation of punching shear reinforcement, use the following check

$$v_u \le v_c \qquad (4.22)$$

   where v_c is the allowable punching shear stress for concrete.

For prestressed two-way slab systems, the allowable punching shear stress of concrete (v_c) is the lesser of the following two equations.[30]

80 Post-tensioning in building construction

$$v_c = 0.083\left(1.5 + \frac{\alpha_s d}{b_0}\right)\lambda\sqrt{f_c'} + 0.3f_{pc} + \frac{V_p}{b_0 d} \quad (4.23)$$

$$v_c = (0.29\lambda\sqrt{f_c'} + 0.3f_{pc}) + \frac{V_p}{b_0 d} \quad (4.24)$$

where

$\lambda = 1$ for normal weight concrete;
f_{pc} = average precompression;
V_p = vertical component of effective prestressing force;
b_o = perimeter of the assumed critical punching section;
d = depth of member to centroid of tension; and
α_s = 40 for interior; 30 for edge; 20 for corner columns.

The effective depth 'd' for punching shear is taken from the bottom of the slab to the distance between the intersecting top reinforcement.

At this preliminary stage, assume the following minimum precompression required by ACI 318.

$V_p = 0$ $f_{pc} = 0.86$ MPa

Punching shear reinforcement is required, if $v_u > v_c$.

If the preceding preliminary punching shear check does not pass, modify the selected parameters of design. The choices for change are the cross-sectional geometry of the support, slab thickness and/or its material properties.

4.8.4 Check for live load deflection

ACI 318 sets the maximum permissible live load deflection[31] as (span/360).

Column-supported floor designs, using ACI-318, are essentially crack-free. Uncracked slab deflection under live load is not impacted by the addition of post-tensioning, neither in amount nor in arrangement. Live load deflection is governed by slab thickness, support arrangement and the material properties of the slab.

Due to strict limitation on live load deflection, through preliminary analysis or otherwise, the adequacy of the floor dimensions and concrete strength must be verified before proceeding to detailed design.

4.8.5 Subdivide the slab into design strips

The design is based on the premise that the floor slab is subdivided into design strips in two orthogonal directions. The concept is detailed in Chapter 1 and reference (Aalami, 2014). Chapter 5 of this book offers an application of the scheme.

4.8.6 Select post-tensioning and arrange tendons

Each of the design strips identified in the preceding step must be provided with an adequate amount of post-tensioning to satisfy the minimum precompression and the maximum allowable tensile stresses stipulated in ACI 318. There are other considerations for detailing that are handled at the final stages of design.

4.8.7 Check for average precompression

ACI 318 requires that the average precompression provided by post-tensioning in each of the principal directions be no less than 0.86 MPa.[32]

In common practice, the check for compliance is carried out by estimating the effective force of post-tensioning in each of the design strips and dividing the force by the cross-sectional area of its associated design strip.

More prestressing must be added if the minimum precompression is not satisfied. ACI 318 does not offer an alternative option.

4.8.8 Analyze structure; obtain design values

4.8.8.1 Analysis of structure

One option is to model the floor using finite element-based software to determine the design values. Another option is to use software based on isolated design strips.

4.8.8.2 Extraction of design values

Each design strip is subdivided into design sections. Typically, one design section is at the face of support; and at least one at midspan. For each design section a single representative extreme fiber stress is calculated – one for the top and one for the bottom face of the slab. The representative stresses, also referred to as 'hypothetical' or 'computed' stresses, are viewed to reflect the state of the stress in the design sections they relate to. The hypothetical stresses are used to validate the code compliance and complete the design.

4.8.9 Serviceability check (serviceability limit state – SLS)

The focus of the serviceability check is on deflection and crack control.

4.8.9.1 Deflection control

ACI 318 requires deflection control for three conditions.[33]

A. **Live load deflection**: Deflection under full live load is limited to span/360.[34]

For ACI 318 compliance, live load deflection for post-tensioned floors is calculated using uncracked sections. If the limitation on live load deflection is not met, the member stiffness must be increased. Increase in member depth and/or concrete strength increases slab stiffness.

Increase in post-tensioning force or changes in the profile of post-tensioning tendons do not impact the live load deflection of the uncracked slab.

B. **Long-term deflection**: The allowable limit on the long-term deflection depends on the impact of the projected deflection on the expected performance of the floor. The impact can be visual; sense of discomfort; or malfunction of attachments to the floor. Long-term deflection can be controlled through camber, topping on the structural slab; or false ceiling, depending on the condition.

The limit for visual effects is set at (span/240).[35]

For an estimate of long-term deflection, ACI 318 recommends the long-term deflection to be considered two times the instantaneous deflection.[36]

C. **Damage to nonstructural elements**: Brittle nonstructural members that are installed while long-term deflection is in progress are likely to damage. The anticipated damage depends on the time of installation and the material of the installed member.

ACI 318 recommends to limit the deflection after installation of nonstructural brittle elements to (span/480).

For an estimate of the amount of deflection that is likely to take place after installation of nonstructural brittle materials, ACI 318 recommends the time-line graph[37] reproduced in Figure 4.8 for conventionally reinforced

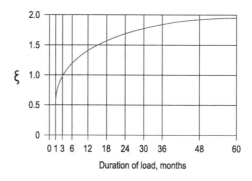

Figure 4.8 Long-term deflection of conventionally reinforced members.

members. Figure 4.8 shows the ACI-recommended estimate guide for long-term deflection of conventionally reinforced concrete members.[38].

4.8.9.2 Load combinations

ACI 318 considers two load combinations: sustained (quasi-permanent) and total (frequent).

The sustained load is the fraction of the live load that is on the member long enough to result in long-term effects, such as creep. It relates to the common occupancy of a floor.

The total (frequent) is the design load. This is the load that, when factored, is used in the safety evaluation of the member. It represents infrequent but probable load application.

ACI 318 leaves it to the judgment of the design engineer to determine the fraction of the design live load to consider as sustained for specific application.

For common residential and office floors, the fraction is commonly assumed 0.3. Hence the applicable load combinations will be:
Sustained condition

$$U = 1.0DL + 0.3LL + 1.0PT \qquad (4.25)$$

Total load condition:

$$U = 1.0DL + 1.0LL + 1.0PT \qquad (4.26)$$

4.8.9.3 Stress checks

ACI 318 sets limits for the hypothetical extreme fiber tension and compression stresses.

Maximum allowable hypothetical tension stress f_t:[39]

$$f_t = 0.5\sqrt{f'_c} \qquad (4.27)$$

Maximum allowable hypothetical compression stress f_c:[40]
For sustained load combination:

$$f_c = 0.45 f'_c \qquad (4.28)$$

For total load combination:

$$f_c = 0.60 f'_c \qquad (4.29)$$

ACI 318 does not offer remedy if the hypothetical tensile stress exceeds the set limit. The parameters of the design must be changed. Other major codes[41] offer crack control options, such as adding measured non-prestressed reinforcement.

4.8.9.4 Minimum non-prestressed reinforcement

ACI 318 requires a minimum area of non-prestressed reinforcement to be included in detailing of post-tensioned column-supported floors. The amount depends on several considerations. The flow chart of Figure 4.9 details the required amount and location of the non-prestressed reinforcement for column-supported post-tensioned floors.

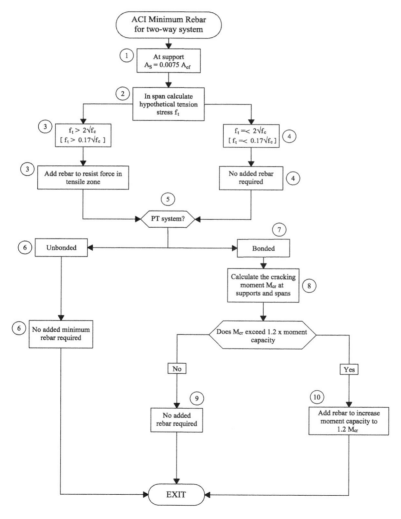

Figure 4.9 ACI 318 minimum non-prestressed reinforcement for two-way slabs.

For ACI 318, the necessity and amount of the minimum non-prestressed reinforcement depends on whether the location is regarded as support or field (span). In this context, the delineation of how far away from the face of support 'span' begins is left to the judgment of the designer. The common practice is to consider the central six-tenth portion of the clear distance between the faces of the adjacent supports as 'span' when checking for minimum rebar.

The following numbers in parentheses refer to the options in the preceding flow chart.

(1) Rebar over the supports: Typically, this is the location of negative moment and precompressed tensile zone. Regardless of the design values and geometry, a minimum amount of bonded reinforcement is required over each support of all two-way floor systems.

This minimum requirement applies to both bonded and unbonded post-tensioning systems.[42]

The area of minimum non-prestressed reinforcement is based on the cross-sectional area of the design strip in direction of analysis and the design strip normal to it[43] as given below.

$$A_{s,\min} = 0.00075 A_{cf} \qquad (4.30)$$

where

$A_{s,min}$ = minimum area of bonded reinforcement; and
A_{cf} = larger cross-sectional area of the design strips of the two orthogonal directions intersecting at the support under consideration.

Figure 4.10 shows the reference locations for the computation of minimum top reinforcement over the supports. For slabs reinforced with bonded tendons, it is permitted to reduce the amount of $A_{s,min}$ by

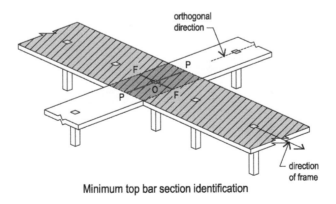

Minimum top bar section identification

Figure 4.10 Reference sections for minimum bonded reinforcement.

the cross-sectional area of bonded tendons that are located within the effective support width.[44]

The larger A_{cf} of the two directions applies. Consequently, the area of minimum bonded reinforcement over support will be the same in both directions.

For typical applications, the effective support width equals the width of the column normal to the span plus 1.5 times the slab thickness on each side of it. Figure 4.11 shows the effective support width for common conditions.

(1) Calculate the hypothetical tensile stress (f_t) in the span: The necessity of the minimum amount of non-prestressed reinforcement in the span depends on the value of the hypothetical extreme fiber tensile stress.
(2) Stress does not exceed the threshold: If the hypothetical tensile stress in the span does not exceed the threshold specified in the following equation, minimum bonded reinforcement is not required in the respective span.

$$f_t \leq 0.17\sqrt{f_c'} \tag{4.31}$$

This concludes the provision of minimum non-prestressed reinforcement for floors reinforced with unbonded tendons.

(3) Stress exceeding threshold: If the hypothetical tensile stress in the span exceeds the value specified in the preceding equation, non-prestressed bonded reinforcement is required in the respective span.

The minimum area of bonded reinforcement is given by:

$$A_{s,min} = \frac{N_c}{0.5 f_y} \tag{4.32}$$

where
$A_{s,min}$ = area of bonded reinforcement;

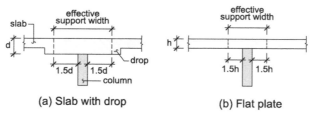

(a) Slab with drop (b) Flat plate

Effective Support Width

Figure 4.11 Effective support width.

f_y = specified yield strength of reinforcement, not to be accounted for in excess of 414 MPa; and

N_c = tension force in the tension zone of concrete at the design section under consideration.

The computation of the tension force N_c using Figure 4.7 is given in the following:

$$N_c = 0.5(h-c)f_t \times b \qquad (4.33)$$

$$h - c = \left[\frac{f_t}{(f_t + f_c)}\right]h \qquad (4.34)$$

$$N_c = 0.5\left(\frac{f_t^2}{f_t + f_c}\right)bh \qquad (4.35)$$

where
b = member width; width of design strip.

(4) Check post-tensioning system whether bonded or unbonded.
(5) Unbonded systems: No additional minimum non-prestressed reinforcement is required. Exit the minimum rebar code check.
(6) Bonded system: The minimum non-prestressed bonded reinforcement in two-way slabs post-tensioned with bonded systems is based on the relationship between their nominal moment[45] capacity and the cracking moment at the same location, as detailed below.
(7) The total amount of prestressed and non-prestressed reinforcement at any section reinforced with bonded tendons shall be adequate to develop nominal moment capacity at that section not less than 1.2 times the cracking moment computed for the same section, using the modulus of rupture of the section. The provision is precaution against abrupt failure that can develop immediately after cracking.

$$M_n \geq 1.2 M_{cr} \qquad (4.36)$$

M_{cr} = cracking moment of the section; and
M_n = nominal moment capacity of the section.

(8) Cracking moment does not exceed the minimum required moment capacity. No added reinforcement is necessary. Exit the minimum rebar check.
(9) Cracking moment exceeds the minimum required capacity. Add reinforcement to meet the requirements.

4.8.10 Safety check (ultimate limit state – ULS)

The strength adequacy of the member against specified overload is checked for two load combinations.

4.8.10.1 Load combinations for strength

The load combinations for strength are:[46]

$$U_1 = 1.20DL + 1.60LL + 1.00HYP \quad (4.37)$$

$$U_2 = 1.40DL + 1.00HYP \quad (4.38)$$

where DL, LL and HYP are respectively dead load, live load and hyperstatic actions from prestressing. If the straight method of design is used, HYP is replaced with PT. The parameter PT in this method includes the hyperstatic effects of post-tensioning.

ACI 318 is specific on the inclusion of hyperstatic effects in the strength design of all post-tensioned members.[47]

4.8.10.2 Strength calculation

The contribution of post-tensioning to flexural strength of a section depends on whether the prestressing steel is bonded or unbonded.

At overload, the stretching results in stress gain in tendons. In lieu of detailed analysis, ACI 318 offers the following simplified equations[48] for gain in tendon stress at ULS.

A. **For members with bonded tendons**: For stress in prestressing steel at ULS f_{ps} the following approximate values are allowed, if the effective stress in prestressing steel (f_{se}) is not less than 0.5 f_{pu} (where f_{pu} is the specified tensile strength of prestressing steel).

$$f_{ps} = f_{pu}\left\{1 - \frac{\gamma_p}{\beta_1}\left[\rho_p \frac{f_{pu}}{f'_c} + \frac{d}{d_p}\frac{f_y}{f'_c}(\rho - \rho')\right]\right\} \quad (4.39)$$

B. **For members with unbonded tendons**: The gain in stress at ultimate limit is related to the respective span.
   (i) For span-to-depth ratio 35 or less:

$$f_{ps} = f_{se} + 70 + \frac{f'_c}{100\rho_p} \quad (4.40)$$

where f_{ps} shall not be greater than f_{py}, nor (f_{se} + 420).
   (ii) For members reinforced with unbonded tendons and span-to-depth ratio greater than 35:

$$f_{ps} = f_{se} + 70 + \frac{f'_c}{300\rho_p} \quad (4.41)$$

where f_{ps} shall not be taken greater than f_{py}, nor greater than (f_{se} + 210) MPa

4.8.11 Punching shear

The evaluation of punching shear, reinforcement for punching shear and detailing for punching shear is given in Chapter 7 of ACI 318.

The recommended procedure does not appear to directly relate or address the punching shear capacity to the amount of tension reinforcement over and around the column. The interested reader is referred to EC2.[49] Reference to ACI 318 punching shear is limited to the maximum allowable value referred to earlier in this chapter.

4.8.12 Initial condition; transfer of prestressing

At transfer of prestressing (jacking), ACI 318 limits the extreme fiber tension and compression stresses.[50]

4.8.12.1 Load combination

ACI 318 does not specify the load combination at transfer of prestressing. The load combination generally used in practice is:

$$U = 1.00SW + 1.15PT \qquad (4.42)$$

where SW is the self-weight of the member at time of stressing.

4.8.12.2 Allowable stresses[51]

For tension:

$$0.25\sqrt{f'_{ci}} \qquad (4.43)$$

Extreme fiber compression:

$$0.6f'_{ci} \qquad (4.44)$$

where f'_{ci} is the compressive strength of 28-day concrete cylinder at transfer of prestressing.

4.8.13 Detailing

4.8.13.1 Tendon arrangement

A. **Integrity steel**: ACI 318 requires that at least two tendons[52] in each of the principal directions pass through the column case (over the support). These tendons must be anchored beyond the face of the column to develop their full strength.

Where it is not practical to pass the specified tendons through the column cage, alternatively non-prestressed reinforcement can be used.[53] This reinforcement shall be anchored to develop its design strength at the face of the column.

B. **Maximum tendon spacing:**[54] At most in one of the principal directions, the tendon spacing shall not exceed 1.5 m, nor eight times the slab thickness.

4.8.13.2 Rebar arrangement

The arrangement of non-prestressed reinforcement and its detailing is demonstrated in the example of Chapter 5.

NOTES

1. EC2 EN 1992-1-1, 2004 (Part 1).
2. ACI 318-19.
3. EC2 Appendix I.
4. The expression 'equivalent frame,' used in EC2 does not refer to the same structural modeling as in ACI 318. The intent of EC2 is a simple frame analysis based on the gross cross-sectional geometry of its members.
5. EC2 Appendix I.
6. ACI 318-11 13.6.4.
7. ACI 318 specifically recommends against subdividing the moment between the column strip and middle strips for post-tensioned floors.
8. The value is recommended in ACI for conventionally reinforced column-supported slabs.
9. Eurocode 1, Part 2.1 (ENV 1991-2-1) EC2 Section 7.3.1.
10. Eurocode 1, Part 2.1 (ENV 1991-2-1) EC2 Section 7.3.1.
11. Eurocode 1, Part 2.1 (ENV 1991-2-1) EC2 Table 7.1N.
12. EC2 Table 3.1
13. Twenty-eight-day cylinder/cube compressive strength in MPa.
14. EC2 Section 7.2.
15. EC2 Section 7.2.
16. EN 1992-1-1:2004(E), Section 9.2.1.1(4)
17. Design Aids for Eurocode 2, part 1 [ENV 1992-1-1], Section 4.1
18. Design Aids for Eurocode 2, part 1 [ENV 1992-1-1], Section 4.1.
19. The recommended value for the live load is given in Section 4.10.1 of EC2. The value selected applies to common building structures.
20. EN 1992-1-1:2004(E), Section 7.2(3).
21. EN 1992-1-1:2004(E), Sections 7.1(2) and 7.3.2(4).
22. EN 1992-1-1:2004(E), Section 4.2, Table 4.1.
23. EC2 EN 1992-1-1:2004(E).
24. EC2 EN 1992-1-1:2004(E), Section 7.3.2(3).
25. EN 1992-1-1:2004(E), Section 7.3.4.

26. EC2 EN 1992-1-1:2004, Section 9.2.1.4
27. ASCE-7 For live load reduction and specification
28. ACI 318-19 22.6.6.2.
29. ACI 318 Table 21.2.1.
30. ACI 318-19 22.6.5.5b; 22.6.5.5a.
31. ACI 318 24.2.2.
32. ACI 318-19 8.6.2.1.
33. ACI 318-19 24.2.2.
34. ACI 318-19 24.2.2.
35. ACI 318-19 24.2.2.
36. ACI 318-19 Table 24.2.4.1.3.
37. ACI 318-19 Fig. R.24.2.1.
38. ACI 319-19 24.2.4.1.3.
39. ACI 318-19 8.3.4.1.
40. ACI 318-19 24.5.4.1.
41. EC2 EN 1992-1-1:2004.
42. ACI 318-19 Table 8.6.2.3.
43. ACI 318-11 Section 18.9.3.3.
44. ACI 318-19 6.3.2.1.
45. ACI 318-19 Section 8.6.2.2.
46. ACI 318-19 Table 5.3.1 and Section 5.3.11.
47. ACI 318-19 Section 5.3.11.
48. ACI 318-19 Section 20.3.2.3.1 through 20.3.2.4.1.
49. EC2 EN 1992-1-1:2004 6.4.
50. ACI 318-19 24.5.3.1; 24.5.3.2.
51. ACI 318-19 24.5.3.1; 24.5.3.2.
52. ACI 318-19 8.7.5.6.1.
53. ACI 318-19 8.7.5.6.3.1.
54. ACI 318-19 8.7.2.3.

REFERENCES

Aalami, B. O. (2021), *Post-Tensioning; Concepts; Design; Construction*, PT Structures, California, www.PTStructures.com, pp. 570.
ACI 318-19 (2019), *Building Code Requirements for Structural Concrete (ACI 318-19) and Commentary*, American Concrete Institute, Farmington Hill, MI, www.concrete.org, 623 pp.
BS8110-1 (1987), *Structural Use of Concrete, Part 1: Code of Practice for Design and Construction*, British Standards Institution, London, UK.
European Code EC2 (2004), *Eurocode 2: Design of Concrete Structures – Part 1-1 General Rules and Rules for Buildings*, European Standard EN 1992-1-1:2004, CEN Brussells.
TR43 (2005), *Post-Tensioned Concrete Floors Design Handbook*, The Concrete Society, Camberley, Surrey, UK, www.concrete.org.uk, 110 pp.
Widianto, O. B. (2009), 'Two-Way Shear Strength of Slab-Column Connections; Reexamination of ACI 318 Provisions,' *ACI Structural Journal*, V. 107, January–February 2010, pp. 120–122.

Chapter 5

Column-supported floor example

5.1 COLUMN-SUPPORTED FLOOR

The following is a detailed long-hand design of a column-supported post-tensioned floor slab, typical of residential or commercial buildings. The design is carried out side-by-side using the European code EC2[1] and the American building code ACI 318.[2]

The example uses the unbonded post-tensioning system. Where appropriate the difference between the selected unbonded system and the bonded alternative is explained. Figure 5.1 is an example of construction using flat slab floors and an unbonded post-tensioning system.

5.2 GEOMETRY; LOAD PATH; DESIGN STRIP

5.2.1 Structure

The floor slab is a typical level of a multi-story concrete frame. Figure 5.2 shows the partial elevation of the concrete frame. The slab is 220 mm thick. It is supported on square columns.

5.2.2 Design strip

For design, the floor slab is subdivided into strips running in both directions. The design strips follow the line of supports. Each strip is extracted and designed in isolation along with its own supports and loading.

Figure 5.3 illustrates a typical design strip of the floor. On each side of the line of support, the design strip extends to the tributary of the support line columns.

The identified design strip is extracted and modeled in isolation with a row of columns above and below. The boundary conditions of the columns at the far ends are assumed rotationally fixed. For typical levels of concrete

94 Post-tensioning in building construction

Figure 5.1 Post-tensioned flat slab building under construction using unbonded tendons.

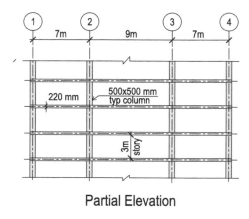

Figure 5.2 Partial elevation of the building frame.

frames, the far ends of the columns in the strip model are modeled as roller support. This results in the dispersion of precompression through the design strip and allows for the shortening of the slab under precompression.

The first elevated slab above the foundation and the roof slab of the concrete frame are likely to lose some prestressing force to the foundation in the former case, and to the level below the roof in the latter.

Figure 5.4 shows the elevation of a typical single level strip selected for design.

5.2.3 Design strip section properties

The section properties of the design strip are listed in Table 5.1. The values are based on the gross cross-sectional geometry of the design strip.

Column-supported floor example 95

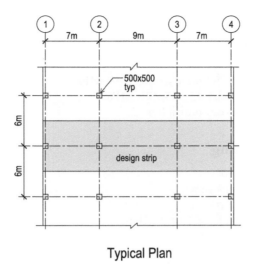

Typical Plan

Figure 5.3 Partial plan of the floor identifying a typical design strip.

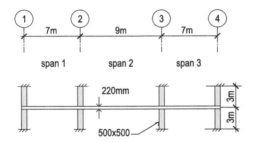

Design Strip Elevation

Figure 5.4 Elevation; typical single-level strip selected for design.

Table 5.1 Section properties of the design strip

Span	Width mm	Thickness mm	Area mm^2	I mm^4	Z mm^3
1	6,000	220	1.32×10^6	5.32×10^9	4.84×10^7
2	6,000	220	1.32×10^6	5.32×10^9	4.84×10^7
3	6,000	200	1.32×10^6	5.32×10^9	4.84×10^7

Note: I is the second moment of area; Z is the section modulus.

5.3 MATERIAL PROPERTIES

5.3.1 Concrete

Weight 24 kN/m³
Twenty-eight-day cylinder strength 40 MPa
Modulus of elasticity E_c = 35,220 MPa
Long-term deflection multiplier[3] 2

5.3.2 Prestressing

System: Unbonded[4]
Strand diameter 13 mm
Strand area 99 mm²
Ultimate strength f_{pk} = 1,860 MPa
Yield stress 1,670 MPa
Jacking force 80% of ultimate strength
Effective stress[5] f_{se} = 1,200 MPa
Modulus of elasticity E_p = 200,000 MPa

5.3.3 Non-prestressed reinforcement

Yield stress f_{yk} [f_y] = 460 MPa
Modulus of elasticity E_s ; E_p = >200,000 MPa

5.4 LOADS

5.4.1 Self-weight

Based on volume; 24 kN/m³

5.4.2 Superimposed dead load

SDL = 2.00 kN/m²

5.4.3 Dead load

DL = 0.2 × 2,400 + 2.00 = 7.28 kN/m²

5.4.4 Live load

LL = 2.50 kN/m²

5.5 DESIGN PARAMETERS

5.5.1 Applicable codes
EC2 EN 1992-1-1; 2004
   ACI 318-19; IBC 2018

5.5.2 Cover to reinforcement
Minimum cover to reinforcement is based on exposure of the member to elements of corrosion and the fire resistivity requirements.

For slabs not exposed to weather, ACI 318-19 recommends 20 mm to prestressing tendon[6] and the same value to non-prestressed reinforcement.[7]

EC2 recommendation is based on exposure classification of the member.[8] Rather than cover, EC2 recommends allowable crack widths – design crack width. For this design example, the same cover is used for both EC2 and ACI options.

Cover to tendon and reinforcement top and bottom is assumed 20 mm. Minimum cover is also required for fire resistivity. For 2-hour fire resistivity, 20 mm applies for interior and 40 mm for exterior (IBC 2021, Table 721.1(1); 4-1.1 and 4-1.2).

5.5.3 Post-tensioning system; effective stress
Unbonded system is selected. For hand calculation, a conservative uniform effective stress is assumed over the entire length of the tendon. The effective stress is the long-term service condition stress in tendon after all losses. The value applicable to unbonded tendons is typically 1,200 MPa. For bonded tendons, the typical value is 1,100 MPa.

5.5.4 Allowable design stress; crack control
For slab construction, depending on the building section of the code followed, crack control is based on either 'maximum allowable' hypothetical tensile stresses,[9] or probable crack widths. ACI 318 imposes maximum limit on the hypothetical tensile stresses. EC2 specifies a threshold for the hypothetical tensile stresses. Using EC2, exceeding the threshold stress triggers an additional check for crack control. Both EC2 and ACI options are followed in the example.

5.5.4.1 EC2 crack control
Depending on the likely exposure of the floor, EC2 recommends 'design crack width w_k.' For the current slab the recommended crack width is w_k = 0.3 mm[10] for quasi-permanent load condition.

Start with frequent load combination given below and calculate the hypothetical tensile stress of the design strip.

Frequent load combination

$$U = 1.00DL + 0.50LL + 1.00PT \tag{5.1}$$

Compare the hypothetical tensile stress with the following threshold.

$$f_{ctm} = 0.3 f_{ck}^{2/3} = 0.3 \times 40^{2/3} = 3.51\,\text{MPa} \tag{5.2}$$

If the hypothetical tensile stress exceeds the above limit, EC2 requires to continue the calculation to determine the probable crack width.

If the calculated crack width exceeds the design crack width (w_k), one option is to add reinforcement to control the probable crack width.[11]

Hypothetical compressive stresses[12] must also be checked against the allowable values. But this does not generally govern the design for common residential and commercial buildings. The check is not carried out in this example.

5.5.4.2 ACI 318 crack control

The extent of probable crack formation in two-way slab systems is checked by limiting the value of the hypothetical tensile stresses of the design strip[13] to $0.5\sqrt{f_c'}$. The same stress limit applies to slabs reinforced with either unbonded or bonded post-tensioning systems.

ACI 318 does not allow the hypothetical tensile stress to exceed the allowable. If it does, the design must be revised.

5.5.5 Fraction of dead load to balance; minimum precompression

For most floor slabs, it is economical to balance 60%–80% of the dead load of the floor by post-tensioning in the span with the largest moment demand. The objective is to minimize the total amount of reinforcement used in design. Other spans need not be balanced to the same extent.

For the current case, the interior span is likely to govern the design. Assume balancing 70% of its dead load. Smaller uplift, such as 60% of dead load, is considered for the shorter exterior spans.

Target uplift from post-tensioning for interior span: $0.7DL = 0.7 \times 7.28 = 5.10\,\text{kN/m}^2$.

Typically, the tendon selected for the critical span is extended through the entire length of the design strip. However, its drape at non-critical spans is reduced to avoid overbalancing. This is commonly achieved by raising the low point of the tendon in the non-governing spans.

In the current configuration, downward force on the end spans can be beneficial to the adjacent longer critical span. For this reason, the tendon profile of the end spans will be adjusted to provide uplift for only 60% of the dead load of the span. This is less than the 70% selected for the interior span.

Uplift for the end spans: $0.60 \times 7.28 = 4.37 \text{ kN/m}^2$.

EC2 does not require minimum post-tensioning in a member. EC2 design objectives can be achieved by any combination of prestressed and non-prestressed reinforcement. Hence, the selection of the preceding amount of post-tensioning will satisfy the EC2 requirements with respect to the amount of prestressing.

ACI requires that the average precompression provided by post-tensioning in a two-way floor slab not be less than 0.86 MPa.[14] This necessitates that before proceeding further in design, the adequacy of the post-tensioning selected be verified for its compliance with the minimum average precompression. Hence, for compliance with the minimum precompression the tendon force and its profile will be reviewed next.

5.5.6 Tendon selection and layout

At construction, tendons follow a smooth curve as shown in the example of Figure 5.5 (a). For hand calculation, however, tendons are assumed to be in the shape of a simple parabola (part b of the figure). In addition, force along the tendon is assumed constant and equal to the 'effective force.'

Simple parabola tendons exert essentially uniform lateral force along their length.[15] The uniform upward force distribution from post-tensioning simplifies the hand calculation.

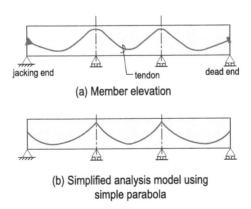

Figure 5.5 Idealization of tendon profile for hand calculation.

Other considerations for mapping the tendon profile in the outline of the concrete member are:
Cover top and bottom 20 mm

CGS[16] top and bottom = 20 + 0.5×13 = 26.5 mm

A. **Select tendon profile for interior span:**
Drape: $a = 220 - 2\times26.5 = 167$ mm
Uplift required: $w_p = 0.70 \times 7.28 = 5.10$ kN/m²
Post-tensioning force P is given by:
$P = w_p L^2/8a = 5.10 \times 9^2/ (8\times0.167) = 309.21$ kN/m width of design strip.
Check the ACI requirement for minimum average precompression.
Average precompression = P/A = 309.21×1,000/(1,000 × 220) = 1.40 MPa > 0.86 MPa OK

B. **Select tendon profile for exterior spans:** The tendons selected for the interior span will be extended through the adjacent spans. The tendon profile in the exterior spans will be adjusted to provide uniform upward force equal to 60% of the dead load of the span. The tendon geometry that provides uniform uplift within the constraints of the end span is shown in Figure 5.6.

Using (i) the geometry constraints shown in Figure 5.6, (ii) the uplift w_p, and (iii) the tendon force P from the adjacent span, the following relationships apply (Aalami, B. O, 2018):

$$d = \frac{\dfrac{2P}{w_p}(e-f)+L^2}{2L} \tag{5.3}$$

$$b = \frac{w_p d^2}{2p} \tag{5.4}$$

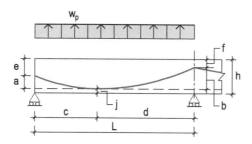

Figure 5.6 Geometry of asymmetrical continuous parabolic tendon for uniform uplift.

Substituting for $P = 309.21$ kN; $w_p = 4.37$ kN/m; $(e - f) = 110 - 26 = 84$ mm; and $L = 7$ m, it gives

$d = 4.34$ m
$c = 7 - 4.34 = 2.66$ m
$b = 4.37 \times 4.34^2/2 \times 309.21 = 0.133$ m
$j = h - (f + b) = 220 - (26 + 134) = 60$ mm > 40 mm from 5.5.2 OK

5.6 SUMMARY OF SERVICE LOADS

Figure 5.7 summarizes the service loads on the design strip consisting of dead, live and prestressing.

5.7 ANALYSIS

With the geometry, boundary conditions, material properties and loads on the design strip defined, the information is complete to calculate the resulting deformations and actions. Common frame analysis or finite element computer programs can be used for the computations. The options are:

(i) Simple frame method (SFM), where the stiffness of the design strip is defined by its gross cross-sectional geometry; and

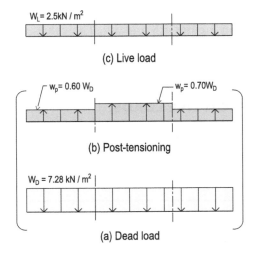

Dead, Live and Post-tensioning
Loads on the Frame

Figure 5.7 Dead, live and post-tensioning loads on the design strip.

(ii) Software based on finite element formulation.[17]

The solutions obtained from the above options will not be the same, leading to different deflections, different stress values, and different amount of reinforcement. However, they are both 'safe' in resisting the specified load. Each alternative satisfies the equilibrium. Each determines the required reinforcement according to its respective computed design values. Also, following the codes, the design sections will have the required ductility.

For the current example. a computer program that is based on SFM is used to determine the distribution of design values in the member.

In summary, the solution is obtained using the gross cross-sectional geometry, the boundary conditions shown in Figure 5.4, along with the loads shown in Figure 5.7.

Live loads must be skipped (patterned) in order to maximize the force demand on the structure. However, based on ACI 318-14,[18] if the magnitude of the live load does not exceed 75% of the dead load, it is not required to skip the live load. The design can be based on the full value of the live load acting over the entire spans of the frame. In this case the ratio of live to dead load is (2.5/7.28 = 0.34 < 0.75). Hence, live load is not skipped. For design, the full value of live load is applied over the entire structure.

5.8 ACTIONS FROM DEAD LOADS

From the analysis of the design strip, the deflection of the slab and the distribution of moments for the dead load on the structure, including the self-weight, are shown in Figures 5.8 and 5.9. The values to be used for hand calculation are those at the face of support and midspans.[19] These are listed in Table 5.2.

5.9 ACTIONS FROM LIVE LOADS

Live load is considered over the entire three spans. Similar to the dead load, the deflection and moments from the application of live load are extracted from the analysis results. The distribution is similar to that of dead load,

Figure 5.8 Instantaneous deflection from dead load; maximum deflection 5.3 mm; deflection at exterior span 2.4 mm.

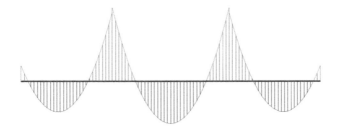

Figure 5.9 Distribution of dead load moment.

Table 5.2 Dead load moments at face of supports and midspan (kNm)

Span	Moment Left	Moment Midspan	Moment Right
1	−29.66	106.88	−217.92
2	−233.62	160.85	−233.62
3	−217.92	106.88	−29.66

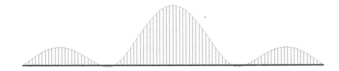

Figure 5.10 Distribution of instantaneous deflection from prestressing.

with the difference that the maximum deflection is at the interior span equal to 1.8 mm, and shown in Figures 5.8 and 5.11 and Table 5.3.

Table 5.3 lists the calculated moments at design locations.

5.10 ACTIONS FROM PRESTRESSING FORCES

Prestressing is designed to counteract the gravity loads. The distribution of deflection from prestressing and prestressing moments is shown in Figures 5.10 and 5.11. The values at the critical points are listed in Table 5.4.

Maximum deflection from prestressing is 3.9 mm upward in central span; deflection at center of exterior spans −1.2 mm.

In addition to the moments from prestressing, design requires the value of the hyperstatic moments. The distribution of hyperstatic moments is shown in Figure 5.12 with the leading numerical values in Table 5.5.

The computation of hyperstatic moments using the universal method is given at the end of this chapter.

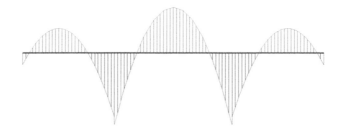

Figure 5.11 Distribution of moment from prestressing. Moment values are given in Table 5.4.

Table 5.3 Live load moments at face of supports and midspan (kNm)

Span	Moment Left	Moment Midspan	Moment Right
1	−10.18	36.70	−74.83
2	−80.21	55.23	−80.21
3	−74.83	36.70	−10.18

Table 5.4 Post-tensioning moments at face of supports and midspan (kNm)

Span	Moment Left	Moment Midspan	Moment Right
1	15.74	−58.17	143.98
2	159.06	−118.59	159.06
3	143.98	−58.17	15.74

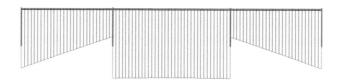

Figure 5.12 Distribution of hyperstatic moments from prestressing. Values of moments are given in Table 5.5.

Table 5.5 Hyperstatic moments from prestressing at face of supports and midspan (kNm)

Span	Moment Left	Moment Midspan	Moment Right
1	31.32	23.72	16.15
2	36.16	36.16	36.16
3	16.15	23.72	31.32

5.11 SERVICEABILITY CHECK; SERVICEABILITY LIMIT STATE (SLS)

5.11.1 EC2 serviceability check

For common residential and commercial building floors, the serviceability check of EC2 is somewhat straightforward. The focus of EC2 serviceability check is crack control and deflection. In general, depending on the environment and application, limited cracking is permitted.[20] Or, if the consequence of cracking is acceptable, the limitation on crack width is relaxed.[21]

Likewise, the limit on computed deflection depends on the consequence of the predicted deflection. Limits on maximum allowable deflection are not imposed, if the immediate and long-term impact of deflection for specific application is acceptable.[22]

The steps necessary to follow for the serviceability check of EC2 are summarized in the EC2 serviceability check flow chart of Chapter 4.

5.11.1.1 Stress thresholds

The stress threshold for the hypothetical tensile stress is f_{ctm}:

$$f_{ctm} = 0.3 f_{ck}^{2/3} \tag{5.5}$$

f_{ck} = 40 MPa
f_{ctm} = 0.3×40$^{2/3}$ = 3.51 MPa tension

Threshold for hypothetical compressive stress for frequent load combination is $0.6f_{ck}$
$0.6f_{ck}$ = 0.6×40 = 24 MPa
Threshold for hypothetical compressive stress for frequent load combination is $0.45f_{ck}$
$0.45f_{ck}$ = 0.45×40 = 18 MPa
The hypothetical extreme fiber stresses will be multiplied by 1.30 to account for the difference between the hypothetical extreme fibers determined based on design strip and middle strip column strip concept.[23]

5.11.1.2 Minimum and maximum reinforcement

There are two requirements for minimum reinforcement. One is for crack control due to shrinkage and temperature.[24] The other provides safety against collapse at initiation of first crack in a section. The latter is dealt with in the safety section of the design. The following covers the minimum reinforcement for shrinkage and temperature control.

A. **Minimum reinforcement**: The required minimum reinforcement (A_{smin}) is

$$A_{smin} \geq \frac{0.26 b_t d f_{ctm}}{f_{yk}} \geq 0.0013 b_t d \tag{5.6}$$

No non-prestressed reinforcement is specified at this stage of design, hence
$A_s = 0$

There is 1.40 MPa precompression from post-tensioning. Considering that the non-prestressed rebar designated for shrinkage control is generally designed at $(3/8)f_y$, the equivalent precompression for the required rebar is:

Average tension provided by shrinkage rebar
$0.001,3 \times (3/8) \times 420 = 0.20$ MPa

Precompression provided by tendons is 1.40 MPa > 0.20 MPa OK

Hence no rebar based on cross-sectional area is required. The compression provided by prestressing exceeds the amount that can be supplied by the prescribed minimum reinforcement.

B. **Maximum reinforcement**: This check[25] verifies that the post-tensioning selected does not over-reinforce the section.

$$A_{smax} = 0.04 A_c \tag{5.7}$$

At this stage the reinforcement is limited to the 14-strand post-tensioning provided (A_{ps})
$A_{ps} = 14 \times 99 = 1{,}386$ mm²

$$A_{sprov} = A_s + A_{ps} \times \frac{f_{pk}}{f_{yk}} \tag{5.8}$$

$A_{sprov} = 0 + 1{,}386 \, (1{,}860/460) = 5{,}604$ mm²
$0.04 A_c = 0.04 \times 220 \times 6{,}000 = 52{,}800$ mm² >> 5,604 mm² OK

The post-tensioning provided does not exceed the maximum recommended value.

5.11.1.3 Check hypothetical extreme fiber stresses for frequent load combination

The frequent load combination is:

$$U = 1.0 DL + 0.5 LL + 1.0 PT \tag{5.9}$$

The extreme fiber hypothetical stress for the design strip is given by:

$$f = (M_d + 0.5 \, M_l + M_{pt})/S + P/A \tag{5.10}$$

where S is the section modulus of the tributary.

The extreme fiber tensile stresses are compared with the code specified threshold. If computed stresses exceed, further checks are required for crack control.[26]

Compression extreme fiber stresses are also checked against the code-specified threshold. If computed values exceed, deflection calculations should allow for impact of creep.[27]

Threshold for hypothetical tensile stress from 5.11.1.1 is:

f_{ctm} = 3.51 MPa tension

The threshold for hypothetical compression stress from 5.11.1.1 is:

$0.6f_{ck}$ = 24 MPa

The hypothetical extreme fiber stresses will be multiplied by 1.30 to account for the difference between the hypothetical extreme fibers determined based on full width of design strip and design strip/column strip concept.[28]

Review of the results indicates that the stress at the face of support of the second span is the most critical condition.

The stresses are checked for frequent load combination. Once calculated, the bending component of the stress is multiplied by 1.30.

M_d = −233.62 kNm
M_l = −80.21 kNm
M_{pt} = 159.06 kNm
P = 1,844.22 kN

A. Check hypothetical extreme fiber for tension stresses
f = [(−233.62 − 0.5×80.21 + 159.06)10^6/(4.84×10^7)]×1.30 + 1,844.22×1,000/1.32×10^6 =
= −1.68 MPa tension < 3.51 MPa OK

B. Check hypothetical extreme fiber for compression Ssresses
The stress is checked for second span, left support, bottom fiber.
f = [(+233.62 + 0.5×80.21 − 159.06)10^6/(4.84×10^7)]×1.30 + 1,844.22×1,000/1.32×10^6 =
= 4.48 MPa compression < 24 MPa OK

5.11.1.4 Check hypothetical extreme fiber compression stresses for quasi-permanent load combination

The load combination is:

$U = 1.0DL + 0.3LL + 1.0PT$ (5.11)

The stress is checked for second span, left support, bottom fiber.

$f = [(+233.62 + 0.3\times80.21 - 159.06)10^6/(4.84\times10^7)]\times1.30 +$
   $1{,}844.22\times1{,}000/1.32\times10^6 =$
   $= 4.04$ MPa compression < 18MPa OK

5.11.1.5 Deflection check

The computed deflection is evaluated against its consequence on the anticipated function of the slab. The recommended values for maximum deflection in relation to the span length are:

A – Deflection for visual effects; allowable values[29] 1/250
B – Live load deflection[30] 1/500

The limit 1/500 entered for live load deflection is EC2's recommendation for the increment of deflection subsequent to when the member is placed in service. EC2 is not explicit on the instantaneous deflection from live load. In practice 1/500 ratio is used.

When considering the visual impact of deflection, and also its impact on the function and cracking of nonstructural brittle installations, the long-term effects of deflection apply. EC2 neither details nor suggests a procedure for estimating the long-term deflection. It is left to the discretion of the design engineer. The procedure recommended in ACI 318 and described in Section 5.11.2.1 is followed.

5.11.1.6 Crack control

Additional computations and provision for added reinforcement apply if the hypothetical extreme fiber tensile stresses exceed the threshold. In this example they do not. Hence, additional computation for crack control does not apply.[31]

5.11.2 ACI serviceability check

The serviceability check consists of deflection check and stress check for crack control. The requirement for minimum reinforcement is also handled at this stage.

5.11.2.1 Deflection check

Deflection check is done first. If not compliant, the member size and/or the material properties of the project must be adjusted to bring the deflection to compliance. The criteria for acceptance for EC2 and ACI 318 are similar:[32]

   (i) Immediate deflection under live load (span/360).
   (ii) For visual effects and sense of comfort (span/240).

(iii) Long-term deflection subsequent to installation of nonstructural elements likely to be damaged (span/480) provided no measures are taken to abate the consequence of deflection.

The deflection check carried out in Section 5.11.1.1 also applies to ACI 318. Hence, it is not repeated here.

A. **Live load deflection**: Live load deflection is not impacted by the amount and disposition of post-tensioning. This is on the assumption that the slab is essentially crack free. Again, for a crack-free slab the amount of prestressing does not impact the live load deflection.

If live load deflection exceeds the limit, the design must be changed. Options are: increase in slab thickness or increase in concrete strength. Each increases the slab stiffness in bending, thereby reduction in deflection.

From the solution

$\delta_l = 1.8$ mm

$\delta_l/L = 1.8/(9,000) = 1/5,000 < 1/360$ OK

B. **Deflection for visual effects**: This is governed by long-term deflection and finish of the structural slab. The assumption for this design example is that the floor is not cambered, nor is it finished with fill, such as mortar on top of the structural slab before the installation of floor covering. Hence, the visual impact is based on the exposed structural slab.

For estimate of long-term deflection, the quasi-permanent load case based on EC2 is used. ACI318 leaves the selection of the fraction of live load that can be considered quasi-permanent 'sustained' to the design engineer. EC2 recommends 0.3 for residential and commercial occupancy. EC2 value is used.

The long-term deflection multiplier due to creep and shrinkage is 2.[33] For sustained load combination, it is assumed that 30% of the design live load will be acting on the structure. Hence the total long-term deflection upon application of the entire live load will be:

$$\delta = (1+2)(\delta_d + \delta_{pt}) + 0.3(1+2)\delta_l + 0.7\delta_l \quad (5.12)$$

where
δ = total long-term deflection;
δ_d = dead load defection;
δ_l = live load deflection; and
δ_{pt} = deflection from prestressing.
Substituting the values:
$\delta = (1 + 2)(5.3 - 3.9) + 0.3(1 + 2)1.8 + 0.7 \times 1.8 = 7.1$ mm
$\delta/L = 7.1/(9,000) = 1/1,268 < 1/240$

C. **Deflection to avoid damage to nonstructural brittle installations**: The first two requirements are checked first. The third is checked if information on the finish of the structure and estimated time for the installation of brittle nonstructural elements is available.

ACI 318[34] provides (Figure 4.8) a guideline for estimating the residual deflection that is likely to take place subsequent to load application.

Deflection check for avoiding damage to nonstructural brittle elements requires additional information. It requires the age of concrete at installation of brittle elements. This, together with the final long-term deflection of the floor provide an estimate of the fraction of the deflection that is likely to take place after the installation of brittle elements. In the absence of detailed information regarding the construction schedule, assume conservatively the installation to take place 30 days after the slab is cast.

Using ACI 318's guideline for day 30, the balance of long-term deflection subsequent to installation of the nonstructural brittle material is approximately 0.25 times the instantaneous deflection. Hence

$$\delta = 0.25(\delta_d + \delta_{pt}) + 1.0\delta_l \qquad (5.13)$$

$= 0.25(5.3 - 3.9) + 1.0 \times 1.8 = 2.15$ mm

$\delta/L = 2.15/9{,}000 = 1/4{,}186 \ll 1/360$ for ACI; or $1/350$ for EC2

Evaluate the calculated deflections with engineering judgment. First, the calculated deflections from the currently common methods vary by over 30% (Aalami, B. O. 2014), depending on the method used. Second, there is no close correction between the computed and measured deflections. Measured deflections often are larger than calculated, using the common calculation methods. Third, estimating the likelihood of damage to nonstructural elements from given floor displacement is often subject to inaccurate assumptions.

5.11.2.2 ACI stress check

For two-way slabs, ACI 318 limits the hypothetical extreme fiber tensile stress to $0.5\sqrt{f_c'}$. This limit applies to both the support and span regions. In hand calculations, the stress check is carried out at the face of support and at midspan.

Commercially available software typically checks the stresses at more locations along the span.

In the current example, the review of the moment diagrams concludes that the critical locations are likely to be the top stress at the face of support of the second span, and the bottom stress at the middle of the same span. The following is the stress check at these locations.

Allowable extreme fiber hypothetical tensile stress for all load combinations:[35]

$$f = 0.5\sqrt{40} = 3.16 \text{ MPa}$$

Allowable extreme fiber hypothetical compressive stresses are:
For total (frequent) load combination

$$f = 0.6f'_c = 0.6 \times 40 = 24 \text{ MPa}$$

For sustained (quasi-permanent) load combination

$$f = 0.45f'_c = 0.45 \times 40 = 18 \text{ MPa}$$

Load combinations for service stress check are not specified in ACI 318. The following is commonly assumed.
For total load (characteristic) combination

$$U = 1.0 \, DL + 1.0 \, LL + 1.0 \, PT \tag{5.14}$$

For sustained (quasi permanent) load combination

$$U = 1.0 \, DL + 0.3 \, LL + 1.0 \, PT \tag{5.15}$$

The fraction of live load used in the sustained load combination is the same as the 'quasi-permanent' coefficient for a similar application.

The extreme fiber hypothetical stress for the design strip is given by:

$$f = (M_d + M_l + M_{pt})/S + P/A \tag{5.16}$$

where S is the section modulus of the tributary.

A. **Stress at face of support of second span**: Total load combination; second span; face of support; top stress:

$M_d = -233.62$ kNm

$M_l = -80.21$ kNm

$M_{pt} = 159.06$ kNm

$P = 1,844.22$ kN

$f = (-233.62 - 80.21 + 159.06)10^6/(4.84 \times 10^7) + 1,844.22 \times 1,000/1.32 \times 10^6$

$= 1.80$ MPa tension < 3.16 MPa OK

Bottom stress; compression
$f = -(-233.62 - 80.21 + 159.06)10^6/(4.84 \times 10^7) + (1,844.22 \times 1,000/1.32 \times 10^6)$
$= +4.6$ MPa compression <24 MPa OK

Sustained load combination; second span; face of support; top stress; tension

The tension stress will be less than the 'total' load condition. Hence, with the same allowable stress value, stress check is not required.

Bottom compression stress; compression
$f = -(-233.62 - 0.3 \times 80.21 + 159.06)10^6/(4.84 \times 10^7) + (1,844.22 \times 1,000/1.32 \times 10^6) = +3.434$ MPa compression <18 MPa OK

B. **Stress at bottom; middle of second span**

$M_d = 160.85$ kNm

$M_l = 55.23$ kNm

$M_{pt} = -118.59$ kNm

$P = 1,844.22$ kN

Total load combination check; bottom tension stress check:
$f = (160.85 + 55.23 - 118.59)10^6/(4.84 \times 10^7) - (1,844.22 \times 1,000/1.32 \times 10^6)$
$= 0.61$ MPa tension < 3.16 MPa OK

Total load combination; top compression stress check:
$f = -(160.85 + 55.23 - 118.59)10^6/(4.84 \times 10^7) - (1,844.22 \times 1,000/1.32 \times 10^6) = -3.41$ MPa compression < 24 MPa OK

By inspection, the compressive stress for sustained load combination will also be within the allowable margin.

C. **Stress at other locations**: From review of the results, the stresses at other locations are less critical than the locations investigated. Hence, stress check is not pursued.

5.11.2.3 ACI minimum rebar

ACI 318 requires minimum bonded reinforcement for post-tensioned floors. The minimum amounts are specified for the region over the column support and possibly in the span.

A. **Minimum rebar based on geometry of the section**: ACI 318 requires minimum area of bonded reinforcement to be placed over the supports.[36] The minimum area is expressed in terms of the cross-sectional geometry of the design strip, and the design strip orthogonal to it. The required area includes bonded tendons, if available. $A_{s,min}$ is given by:

$$A_{s,min} = 0.000,75 A_{cf} \tag{5.17}$$

A_{cf} is the larger of the gross cross-sectional areas of the design strips in the two orthogonal directions of the support under consideration.

The width of the design strip in the direction of analysis is 6 m. The width of the strip in the orthogonal direction is 8 m. The 8 m width governs. Hence

$A_{s,min} = 0.000{,}75 \times 8{,}000 \times 220 = 1{,}320$ mm²

Use 12 times 12 mm bars

Provided rebar $A_{s,prov}$ is:

$A_{s,prov} = 12 \times 113 = 1{,}356 > 1{,}320$ mm² OK

Place the bars over the support in each direction. Follow the detailing recommendation in Section 5.15.

B. **Minimum rebar based on the level of computed stresses:** The minimum area of non-prestressed reinforcement in the span depends on the value of the hypothetical extreme fiber tensile stress of the design strip over its central span length.[37]

Where the hypothetical tensile stress exceeds $(0.167\sqrt{f_c'})$ it is required to provide bonded reinforcement. The minimum area of the bonded reinforcement depends on the tension force N_c at the location of stress check. The force N_c is shown in Figure 5.13. The area of the required reinforcement A_s is given by:

$$A_s = N_c/0.5f_y \qquad (5.18)$$

The values of the extreme fiber tensile stresses for the total load combinations are:

Span 1: $f = 0.78$ MPa $< 0.167\,f_c'^{0.5} = 1.05$ MPa OK
Span 2: $f = 0.61$ MPa $< 0.167\,f_c'^{0.5} = 1.05$ MPa OK

At both locations, the tensile stresses are below the threshold that triggers added bonded reinforcement.

However, where required, the value of N_c is given by the following relationship.

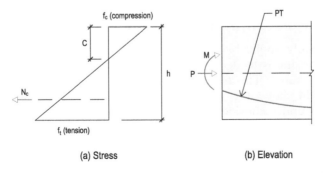

(a) Stress   (b) Elevation

Figure 5.13 Distribution of stress through the depth of slab. N_c is the integral of tension stress shown.

$$N_c = 0.5\left(\frac{f_t^2}{f_t + f_c}\right)bh \qquad (5.19)$$

5.12 SAFETY CHECK; ULTIMATE LIMIT STATE (ULS)

The load combination for the ULS, in addition to contributions from dead and live loads, includes contribution of hyperstatic actions from prestressing. The steps to follow for ULS design are:

(i) Calculate the hyperstatic moments (M_{hyp}).
(ii) Form factored load combination for demand moments (M_u).
(iii) Check capacity of the member against demand moment; add reinforcement, where capacity falls short of demand.
(iv) Check margin of capacity for cracking moment; add reinforcement where needed.
(v) Check for punching shear.

5.12.1 Strength design versus capacity check

For code-required safety check of post-tensioned members, it is expeditious to first calculate the moment capacity of the member using the outcome of the member's serviceability check. Where capacity falls short of demand for strength add non-prestressed reinforcement. This is based on the observation that in most instances the post-tensioning and non-prestressed reinforcement required to meet the serviceability condition provide member strength in excess of the value required to meet the safety check requirement.

In this example, the serviceability check of the member did not conclude with added non-prestressed reinforcement. Hence, the capacity will be based on only the post-tensioning from the serviceability check.

5.12.2 Calculate hyperstatic moments

The universally applicable method for calculating the hyperstatic forces from prestressing is to derive them from the member's support reactions from prestressing.[38] This method is used next.

Figure 5.14 shows the support reactions from post-tensioning for the design strip under consideration. The forces and moments shown in the figure are listed in Table 5.6. The reactions from post-tensioning are always in self-equilibrium – they add up to zero. This is because the post-tensioning forces that cause the reactions are themselves in self-equilibrium.

The hyperstatic reactions shown in part (a) of Figure 5.14 result in the distribution of moment shown in Figure 5.12.

Column-supported floor example   115

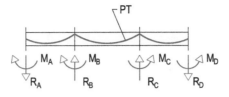

(a) Prestressing reactions

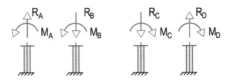

(b) Prestressing reactions

Support Reactions from Prestressing

Figure 5.14 Hyperstatic actions from prestressing at the supports.

Table 5.6 Hyperstatic support reactions from prestressing

Support	A	B	C	D
Reaction kN	2.34	−2.34	−2.34	2.34
Moment kNm	−31.90	−20.59	20.59	31.90

As an example, the following illustrates the calculation of moment at the face of the second support, using the values listed in Table 5.6.

AB = CD = 7 m
BC = 9 m

At support A
M_A = 31.90 kNm

At support B
At left of support M_B = 31.90 − 2.34×7 = 15.52 kNm
At right of support M_B = 15.52 + 20.59 = 36.11 kNm

The values obtained are at the centerline of the support. The moments at the face of support are likely to be somewhat different. Since the variation of hyperstatic moment along a span is linear, the moment at the face of support is obtained using linear interpolation.

For the interior span, the face of support and support center moments will be the same, since the value of moment along this span is constant. For the second support, the moment at the left of the support is given by:

Moment at centerline of first support (A) 31.90 kNm
Moment at centerline of second support (B) 15.52 kNm
Span 7 m
Column width 500 mm
Moment at left of the column face = 15.52 + (0.25/7)(31.90 − 15.52) = 16.10 kNm

The value calculated agrees with the value listed in Table 5.5.

5.12.3 Strength check (ULS)

5.12.3.1 EC2 strength check

A. **Load combinations**: Load combinations for strength check in EC2 depend on the application. The following load combination is used.

$$U_1 = 1.35DL + 1.50LL + 1.00HYP \qquad (5.20)$$

Review of the moment values concludes that the critical demand is at the face of support of the interior span. The values from Tables 5.2–5.4 are:
Dead load moment = −233.62 kNm
Live load moment = −80.21 kNm
Hyperstatic moment = 36.16 kNm
U_1 = −1.35×233.62 −1.50×80.21 + 36.16 = −399.54 kNm

B. **Moment capacity calculation/design**: The moment capacity at the face of the support is calculated using the simplified procedure.[39] The geometry of the section and its reinforcement are shown in Figure 5.15. Total tensile force (T) at ULS is the sum of tension from prestressing (T_p) and tension from rebar (T_s)
$T = T_p + T_s$
Design stress of tendon at ULS = 1,200 + 100 = 1,300 MPa
Design stress of rebar at ULS = 460/1.15 = 400 MPa
There are 16 tendons in the design strip.
T_p = 16×99×1,300/1,000 = 2,059.20 kN

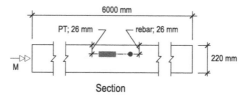

Figure 5.15 Section – slab and position of its reinforcement.

Top rebar 12 times 12 mm bars
$A_s = 12 \times 113 = 1{,}356$ mm²
$T_s = 1{,}356 \times 400/1{,}000 = 542.40$ kN
$T = 2{,}059.20 + 542.40 = 2{,}601.60$ kN
Rectangular block compression stress in the compression zone f_c for
$f_{ck} = 40$ MPa is
$f_c = f_{cd} = 40/1.50 = 26.67$ MPa
Depth of compression zone 'a' is
$a = T/(f_{cd}b) = (2{,}601.60 \times 1{,}000)/(26.67 \times 6{,}000) = 16.26$ mm
Moment capacity is
$M_{rd} = [T_p \times z_p + T_s \times z_s]$
where z_p and z_s are distances from the centroid of the respective tensions to the center of the compression block (a).
$M_{rd} = [2{,}059.20 \times (220-26-0.5 \times 16.26) + 542.40 \times (220-26-0.5 \times 16.26)]/1{,}000 = 483.56$ kNm
Design moment capacity $M_{rd} = 483.56$ kNm > demand moment $M_u = 399.54$ kNm OK
Hence, no added rebar is necessary.

5.12.3.2 ACI strength check

A. **Load combinations**: Two load combinations apply for the gravity design of the floor.[40]

$$U_1 = 1.2DL + 1.6LL + 1.0HYP \qquad (5.21)$$

$$U_2 = 1.4DL + 1.0HYP \qquad (5.22)$$

The second load combination rarely applies to common commercial and residential floors.

Review of the moment values concludes that the critical demand is at the face of support of the interior span. The values from Tables 5.2 and 5.3 are:
Dead load moment = −233.62 kNm
Live load moment = −80.21 kNm
Hyperstatic moment = 36.16 kNm
$U_1 = -1.20 \times 233.62 - 1.60 \times 80.21 + 36.16 = -372.52$ kNm

B. **Moment capacity calculation/design**: The moment capacity at the face of the support is calculated using the simplified, but conservative procedure.

GIVEN
Geometry: The cross-sectional geometry and the position of the reinforcement are given in Figure 5.15.
Width: $b = 6{,}000$ mm
Depth: $h = 220$ mm

Concrete: f'_c = 40 MPa
Prestressing: 16 times 13 mm strands; strand area = 99 mm²; CGS[41] 26 mm from top fiber
Strands stressed to $0.80f_{pu}$; effective stress after losses 1,200 MPa
Distance of center of tension to extreme fiber 26 mm
Non-prestressed reinforcement at top
12 – 12 mm 460 MPa bars; cover 20 mm

REQUIRED
Calculate the design moment capacity of the section, using the simplified procedure.[42]
Total tensile force (T) at ULS is the sum of tension from prestressing (T_p) and tension from rebar (T_s).
$T = T_p + T_s$
Tendon stress at ULS = 1,600 MPa
T_p = 16×99×1,600/1,000 = 2,534.40 kN
A_s = 12×113 = 1,356 mm²
T_s = 1,356×460/1,000 = 623.76 kN
T = 2,534.40 + 623.76 = 3,158.15 kN
Rectangular block compression stress in the compression zone f_c
$f_c = 0.85f'_c = 0.85×40 = 34$ MPa
Depth of compression zone 'a' is
$a = T/(0.85f'_c\, b) = (3,158.15×1,000)/(34×6,000) = 15.5$ mm
Moment capacity φM_n is:

$$\varphi M_n = 0.9[T_p z_p + T_s z_s] \tag{5.23}$$

where z_p and z_s are distances from the centroid of the respective tension forces to the center of the compression block (a).
φM_n = 0.9[2,534.40× (220−26−0.5× 15.5) + 623.76× (220−26−0.5×15.5)]/1,000 = 530 kNm
Design moment capacity φM_n = 530 kNm > demand moment M_u = 372.52 OK
Hence, no added rebar is necessary.

5.12.4 Cracking moment safety check

To avoid abrupt failure, the moment capacity at sections along a member shall be greater than the moment that initiates cracking (M_{cr}) at the same section. This stipulation is included in both EC2 and ACI 318. Its application, however, is different in the two codes (Aalami, B. O., 2016).

5.12.4.1 EC2 cracking moment check

EC2 requires that slabs reinforced with unbonded tendons develop moment capacity 1.15 times the cracking moment.[43]

The equation for cracking moment M_{cr} from Chapter 4 is:

$$M_{cr} = (f_r + P/A)S \qquad (5.24)$$

where f_r is the cracking stress of concrete in bending, and S is the section modulus. For cracking stress f_r EC2 recommends[44]

$$f_r = 0.3 f'_{ck}{}^{2/3} \qquad (5.25)$$

For f'_c = 40 MPa
$f_r = 0.3 \times 40^{2/3}$ = 3.51 MPa
The average precompression of the floor is:
P/A = 1.4 MPa
$S = bd^2/6 = 6{,}000 \times 220^2/6 = 48.4 \times 10^6$ mm^3
M_{cr} = (3.51 + 1.4) ×48.4 ×10^6/10^6 = 237.64 kNm
Required moment capacity M_{rd} = 1.15 × 237.64 = 273.29 kNm

A. **Face of support, second span:** Moment at the face of second span support is:
   U_1 = −1.35×233.62 − 1.50×80.21 + 36.16 = −399.54 kNm > 1.15 M_{cr} = = 273.29 kNm
   At this location, the section is provided with capacity that exceeds the required factored cracking moment. No added rebar is necessary.
B. **Middle of second span**: Moment at the middle of second span is:
   U_1 = −1.35×160.85 − 1.50×55.23 + 36.16 = −263.83 kNm <1.15 M_{cr} = = 273.29 kNm
   The reinforcement for midspan must be slightly increased to cover the shortfall of cracking moment requirement.
   For strict code compliance, the moment capacity along the entire length of the member shall not be less than 1.15M_{cr}, namely 273.29 kNm. This contradicts the common practice, where for the current type of application the reinforcement for cracking moment is checked only at critical locations.[45]

5.12.4.2 ACI cracking moment check

ACI 318 applies the provision to two-way slabs reinforced with bonded tendons.[46] Hence, no additional check is required for the current floor, since it is reinforced with unbonded tendons.

5.13 PUNCHING SHEAR CHECK

Both EC2 and ACI punching shear checks are based on calculating a hypothetical shear stress acting on an assumed critical section and comparing the calculated value with the allowable. If the hypothetical stress exceeds the allowable stress, remedy is called for.

The punching shear values are checked at the second support. The values at the connection of the lower column to the slab are:

Dead load V_d = 378.41 kN downward; M_d = 19.37 kNm
Live load reaction V_l = 129.95 kN downward; M_l = 6.65 kNm
Hyperstatic values V_{hyp} = −2.34 kN upward; M_{hyp} = −20.59 kNm

5.13.1 Based on EC2

A. **Punching shear demand**: The 'frequent' load combination is used.

$$U_1 = 1.35DL + 1.50LL + 1.00HYP \tag{5.26}$$

The design punching shear check is:
V_{Ed} = 1.35×378.41 + 1.50×129.95 − 2.34 = 703.44 kN

The stress check does not include the magnitude of the column moment. Presence or absence of column moment is accounted for through a multiplier for direct shear resulting from column reactions.

The punching shear check/design is carried out at the face of column, and the first critical perimeter shown in Figure 5.16.

The punching shear check will be continued for subsequent assumed sections, until the shear capacity of concrete alone exceeds the demand shear.

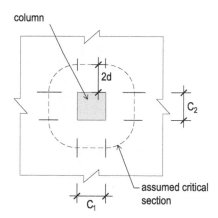

First Punching Shear Critical Section Based on EC2

Figure 5.16 EC2 assumed first punching shear critical section beyond the column perimeter.

B. **Punching shear check at face of support:**
$c_1 = c_2 = 500$ mm
$u_0 = 2(c_1 + c_2) = 2 (500 + 500) = 2,000$ mm
The demand shear stress v_{Ed} is given by:

$$v_{Ed} = \frac{\beta V_{Ed}}{u_0 d} \quad \text{MPA} \tag{5.27}$$

$d = 220 - 20 - 12 = 188$ mm
V_{Ed} = factored demand shear force [N]; and
β [47] = 1.5 for corner;
= 1.15 for interior; and
= 1.4 for end/edge.
For interior columns $\beta = 1.15$
$v_{Ed} = 1.15 \times 703.44 \times 1,000/(2,000 \times 188) = 2.15$ MPa
Maximum allowable shear[48] is $v_{Rd,max}$

$$v_{Rd,max} = 0.5 \eta f_{cd} \tag{5.28}$$

$$\eta = 0.6 \left[1 - \frac{f_{ck}}{250} \right] \tag{5.29}$$

$\eta = 0.6[1 - f_{ck}/250] = 0.6[1 - 40/250] = 0.50$
$v_{Rd,max} = 0.50 \times 0.50 \times f_{cd}$
$f_{cd} = f_{ck}/1.50 = 40/1.5 = 26.67$ MPa
$v_{Rd,max} = 0.5 \times 0.5 \times 26.67 = 6.67$ MPa
$v_{Ed} = 2.15 < v_{Rd,max} = 6.67$ MPa OK

C. **Punching shear check at other assumed critical sections:** Continue to the first assumed critical section (Figure 5.16).
The perimeter of the assumed section is:
Perimeter = $4c + 4d\pi$ = $4 \times 500 + 3.14 \times 4 \times 188 = 4,361$ mm
Area of assumed critical section = $188 \times 4,361 = 819,868$ mm^2
$v_{Ed} = 1.15 \times 703.44 \times 1,000/819,868 = 0.99$ MPa
Calculate the allowable shear stress for concrete[49] $v_{Rd,c}$

$$v_{Rd,c} = C_{Rd,c} k \left(100 \rho_1 f_{ck}\right)^{1/3} + k_1 \sigma_{cp} \geq \left(v_{min} + k_1 \sigma_{cp}\right) \tag{5.30}$$

First check whether the section is adequate without reinforcement, namely $\rho_1 = 0$.
The section is considered adequate if the following is satisfied:

$$v_{Rd,c} \geq \left(v_{min} + k_1 \sigma_{cp}\right) \tag{5.31}$$

where

$$v_{min} = 0.035 k^{3/2} \times f_{ck}^{1/2}; \tag{5.32}$$

$k_1 = 0.1^{50}$; and
σ_{cp} = average of normal stress on control section (MPa, positive if compression).
From design of the slab: $\sigma_{cp} = 1.4$ MPa

$$k = 1 + \sqrt{(200/d)} \leq 2.0, d \text{ value in mm} \tag{5.33}$$

$k = 1 + \sqrt{200/188} = 2.03$, assume 2.0
$v_{min} = 0.035 \times 2^{3/2} \times 40^{1/2} = 0.63$ MPa
$(v_{min} + k_1\sigma_{cp}) = 0.63 + 0.1 \times 1.4 = 0.77$ MPa
$v_{Ed} = 0.99 > 0.77$ NG

Hence, punching shear reinforcement is required.

Unlike ACI 318, where increasing the punching shear strength is by way of adding shear reinforcement, EC2 has the option of adding tension reinforcement over the column.

In the following, the option of adding tension reinforcement over the support is selected.

For impact of tension reinforcement over the support on the punching shear capacity, the following relationship applies:

$$v_{Rd,c} = C_{Rd,c} k (100 \rho_1 f_{ck})^{1/3} \tag{5.34}$$

$$C_{Rd,c} = 0.18 / \gamma_c \tag{5.35}$$

$C_{Rd,c} = 0.18/1.5 = 0.12$
f_{ck} = 28-day cylinder strength of concrete; 40 MPa

$$\rho_1 = \sqrt{\rho_{ly} \times \rho_{lz}} \leq 0.02; \tag{5.36}$$

ρ_{ly}, ρ_{lz} = bonded tension steel in two orthogonal directions over assumed critical section under consideration.

From gravity design of the slab, the specified reinforcement is: 12 times 12 mm bars in each direction. Cross-sectional area of each 12 mm bar is 113 mm².

Width of the assumed critical section is:
$c + 4d = 500 + 4 \times 188 = 1,252$ mm
The specified bars are to be placed at 150 mm on center.
Number of top bars within the width of the assumed critical section is:
Number of bars = 1,252/150 = 8.34, assume 8 bars
ρ_{ly}, ρ_{lz} = 8 × 113/(188×1,252) = 0.003,8

$$\rho_1 = \sqrt{\rho_{ly} \times \rho_{lz}} = 0.0038$$

The contribution of the area of unbonded reinforcement is not accounted for. Area of bonded reinforcement will be considered.

$\rho_{ohcp} = 0$, for unbonded tendons from Equation (5.30).

$$v_{Rd,c} = C_{Rd,c} k \left(100\rho_1 f_{ck}\right)^{1/3} + k_1 \sigma_{cp} \geq \left(v_{min} + k_1 \sigma_{cp}\right)$$

$v_{Rd,c}$ = 0.12×2(100×0.003,8×40)$^{1/3}$ + 0.1×1.4 = 0.73 MPa
v_{Ed} = 0.99 MPa > $v_{Rd,c}$ = 0.73 MPa NG
More reinforcement is required.

Change the 12 mm bars over the assumed critical perimeter to 10 times 18 mm bars.
Area of each 18 mm bar 254 mm²
$\rho_{lx} = \rho_{ly}$ = 10×254/(188×1,252) = 0.010,8

$$\rho_1 = \sqrt{\rho_{ly} \times \rho_{lz}} = 0.0108$$

$v_{Rd,c}$ = 0.12×2(100×0.010,8×40)$^{1/3}$ + 0.1×1.4 = 0.98 MPa
v_{Ed} = 0.99 MPa ≈ v_{Rd} = 0.98 MPa OK
No further check is required.

5.13.2 Based on ACI[51]

A. **Punching shear demand**: The punching shear demand consists of the factored column reaction, along with a fraction of the column moment. While it is recognized that in the general case moments act in both directions, in checking the punching shear stresses, each moment is considered separately together with the column reaction. The calculated stresses at the same location contrary to mechanics of solids in the same direction – contrary to mechanics of solids – are not considered to be additive.[52]

$$V_u = 1.2 V_d + 1.6 V_l + 1.0 V_{hyp} \qquad (5.37)$$

V_u = 1.2×378.41 + 1.6×129.95 −1.0×2.34 = 659.67 kNm
M_u = 1.2×19.37 +1.6×6.65 −1.0×20.59 = 13.29 kNm

A fraction[53] of the factored moment is assumed to be resisted by the hypothetical shear stresses on the assumed critical punching surface. For square columns the fraction is 40%.
Moment to be resisted by punching shear = 0.4 ×13.29 = 5.32 kNm

B. **Punching shear resistance**: The geometry of the column-slab connection for the second support and the details of the first critical section for punching shear check are shown in Figure 5.17.
For prestressed two-way slabs, the allowable punching shear stress of concrete (v_c) is the lesser of the following two equations.[54]

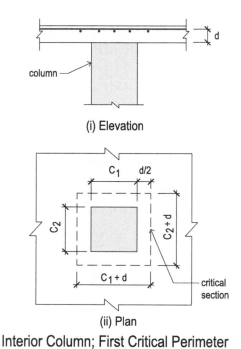

Figure 5.17 General arrangement and parameters of an interior column-slab connection for punching shear check.

$$v_c = 0.083\left(1.5 + \frac{\alpha_s d}{b_0}\right)\lambda\sqrt{f'_c} + 0.3f_{pc} + \frac{V_p}{b_0 d} \tag{5.38}$$

$$v_c = (0.29\lambda\sqrt{f'_c} + 0.3f_{pc}) + \frac{V_p}{b_0 d} \tag{5.39}$$

where
λ = 1 for normal weight concrete;
f_{pc} = average precompression;
V_p = factored punching shear;
b_o = perimeter of the assumed critical punching section;
d = depth of member to centroid of tension; and
α_s = 40 for interior; 30 for edge; 20 for corner columns.
The effective depth 'd' for punching shear is taken from the bottom of the slab to the distance between the intersecting 12 mm top reinforcement.
In common slab construction, the contribution of the term $(V_p/b_0 d)$ is considered small. It is generally disregarded.
$d = 220 - 20 - 12 = 188$ mm

$c_1 + d = c_2 + d = 500 + 188 = 688$ mm
$b_0 = 4 \times 688 = 2{,}752$ mm
$A_c = b_0 \times d = 2{,}752 \times 188 = 517{,}376$ mm^2
Distance of center to first critical perimeter cab
$cab = 688/2 = 344$ mm
λ is 1 for normal weight concrete. α_s is 40 for interior columns.
From Equation (5.38)
$v_c = 0.083\ (1.5 + 40 \times 188/2{,}752) \times 1 \times \sqrt{40} + 0.3 \times 1.4 = 2.65$ MPa
From Equation (5.39)
$v_c = 0.29 \times 1.0 \times \sqrt{40} + 0.3 \times 1.4 = 2.26$ MPa
Assume $v_c = 2.26$ MPa
V_u and M_u are both factored values. They include the hyperstatic contribution from prestressing.
The design shall be based on V_u and $\gamma M_{u,b}$ (40% of column moment by shear[55]).

C. **Punching shear design:** The evaluation of the fictitious stresses over the assumed critical section takes place at the distance from the point of action of the resultant shear to the assumed critical perimeter under consideration. Figure 5.18 shows the distances AB and CD from the centroid.

$$v_{u(AB)} = \frac{V_u}{A_c} + \frac{\gamma_v M_u c_{AB}}{J_c} \qquad (5.40)$$

$$v_{u(CD)} = \frac{V_u}{A_c} + \frac{\gamma_v M_u c_{CD}}{J_c} \qquad (5.41)$$

$$J = \frac{d(c_1+d)^3}{6} + \frac{(c_1+d)d^3}{6} + \frac{d(c_2+d)(c_1+d)^2}{2} \qquad (5.42)$$

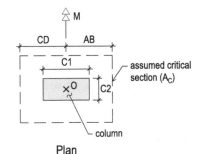

ACI-318 Punching Shear Parameters

Figure 5.18 Plan; definition of farthest distances of the assumed critical perimeter to the center of shear.

$J = 188\times(688)^3/6 + 688\times188^3/6 + 188\times688\times688^2/2 = 4{,}158\times10^{10}$ mm^4

$v_u = 659.67\times10^3/517{,}376 + 0.4\times13.29\times10^6\times344/4{,}158\times10^{10} = 1.28 + 0.04 = 1.32$ MPa $< v_c = 2.26$ MPa

No punching shear reinforcement required.

5.14 INITIAL CONDITION; TRANSFER OF PRESTRESSING

At stressing (i) concrete has not achieved its design strength; (ii) prestressing force is at its highest value; and (iii) live load generally counteracted by prestressing is absent.

Consequently, under the jacking force, the stresses experienced by a post-tensioned member can be significantly different from the member's service condition.

Post-tensioned members are checked for both tension and compression stresses at transfer of prestressing.

Where computed compression stresses exceed the allowable values, stressing is delayed until either concrete gains adequate strength or the member is loaded. Application of design load generally results in lowering the compression stresses present at stressing.

High compression stresses at early age concrete can lead to excessive creep deflection. For this reason, they are controlled.

Where computed tension stresses are excessive, EC2[56]/ACI[57] suggest adding non-prestressed reinforcement at the location of computed high tensile stresses to control cracking.

Higher tension stresses than service condition are tolerated on the premise that at de-shuttering, and the application of load, the tension stresses of the initial condition will be reduced, and possibly turn to compression.

For the current design, the uplift from post-tensioning is less than the self-weight of the slab. For the exterior span post-tensioning balances 60% of the self-weight. For the interior span it balances 70% of the self-weight. Consequently, at stressing, the slab does not lift off its bed. The slab simply shortens under precompression of post-tensioning. This results in uniform compression in the slab at transfer of post-tensioning. Removal of shuttering leads to flexure of the slab.

For illustration of the design steps, as well as the condition when the slab supports are removed, the design steps for the initial condition are continued.

5.14.1 Load combination

The following load combination is generally used in practice for checking the stresses at transfer of prestressing. The combination is neither given in EC2 nor ACI.

$$U = 1.0DL + 0.0LL + 1.15PT \qquad (5.43)$$

In the current example tendons are specified to be stressed when concrete cylinder reaches 30 MPa.[58]
f'_{ci} = 30 MPa

5.14.2 EC2 stress check

A. **Allowable stresses**
   Tension: EC2 is not explicit on allowable tension stresses at stressing.
   Compression 0.6 f_{ck}[59] = 0.6×30 = 18 MPa
B. **Stress check**
   The stress check is performed at the face of the support of the interior span.

$$\sigma = \pm(M_d + 1.15M_{PT})/S + 1.15P/A \tag{5.44}$$

$S = I/Y_c$
Check for compression
σ = (233.62 − 1.15×159.06)(10^6)/4.84×10^7 + 1.15×1.40 = 2.65 MPa
<18 MPa compression OK

5.14.3 ACI stress check

A. **Allowable stresses**[60]
   Tension[61] $0.25\sqrt{f'_{ci}} = 0.25\sqrt{30} = 1.40$ MPa

   Compression[62] 0.6× f'_{ci} = 0.6×30 = 18 MPa
B. **Stress check**
   The stress check is performed at the face of the column of the interior span.
   $\sigma = \pm(M_d + 1.15M_{PT})/S + 1.15\ P/A$ (5.45)
   $S = I/Y_c$
   Strictly speaking, in the preceding equation self-weight moment should be used.
   Check for tension:
   σ = (−233.62 + 1.15×159.06)(10^6)/4.84×10^7 + 1.15×1.40 = 0.56 MPa
   <1.40 MPa tension OK
   Check for compression
   σ = (+233.62 − 1.15×159.06)(10^6)/4.84×10^7 + 1.15×1.40 = 2.65 MPa
   <18 MPa compression OK

5.15 DETAILING

5.15.1 EC2 detailing

EC2 does not offer specific recommendations on detailing of the post-tensioning tendons and reinforcement for flat slab construction.

5.15.2 ACI detailing

General recommendations for tendon layout and non-prestressed trim bars are covered in Chapter 9. The following are items specific to the current design.

- A. **Transfer of column moment:** Where all or a fraction of column moment is resisted by reinforcement in the slab, this reinforcement must be placed at a specified location, namely within a narrow band of slab over the column[63] (effective support width).

    The effective support width is defined[64] as column width plus 1.5 times the slab thickness on each side of the column (Figure 5.17).

    From the section on punching shear check, the factored column moment is determined to be:

    $M_{u,c}$ = 13.34 kNm

    The effective support width is:

    b_e = 500 + 2×1.5×220 = 1,160 mm

    Conservatively, assume that the tendons are grouped normal to the direction of the analysis. Hence, in the direction of analysis the tendons are spaced equally apart. In this tendon arrangement, not all the tendons of the tributary will contribute toward the transfer of column moment.

    As is explained in the next section on 'integrity reinforcement,' at least two tendons pass over the column. Hence, conservatively consider the contribution of two tendons for transfer of column moment.

    The top rebar computed earlier is placed at 150 mm spacing. Hence the number of top bars within the effective support width is:

    Number of top bars: 1,160/150 = 7.70; eight bars were specified for gravity design.

    For the design based on EC2, the top bars are 18 mm. For the design based on ACI 318 the top bars specified are all 12 mm. Conservatively, check the requirement for 12 mm bars.

    Determine the capacity of the section with the following reinforcement: Two unbonded tendons and 8–12 mm bars.

    The position of this reinforcement on the concrete section of the effective support width and the concrete parameters are the same as outlined in Section 5.12.3 for the strength calculation of the entire tributary. The calculation gives the capacity 372.52 kNm.

    The calculated capacity exceeds the demand for transfer of column moment. No added rebar is required.

- B. **Integrity steel:** ACI 318 requires that a minimum amount of reinforcement pass through the column cage.[65] This reinforcement is referred to as 'integrity steel.' The integrity steel shall develop its full tensile force at the face of the column capable of supporting the weight of the slab tributary

to the same column. No calculation is required to determine the amount of the integrity steel, provided the following is met:
(i) Either pass two 13 mm tendons through the column cage in each direction. These tendons shall be anchored beyond the column;
(ii) Or pass two continuous bottom bars from one column to the next.
In both cases the reinforcement has to be anchored to develop its design strength at the face of column.
In the current project, the requirement will be satisfied by specifying a minimum of two strands to pass through the column cage.

5.16 TRIM BARS

In addition to the reinforcement determined for the design to meet the requirements of the building code used, non-prestressed reinforcement is added at selective locations for improved performance. This reinforcement is referred to as steel for 'structural detailing,' or 'trim bars.'

The objective of the trim bars is to avoid/control local cracking and improved distribution of the applied loads. Common trim bars for post-tensioned column-supported floors are given in Chapter 9.

NOTES

1. EC2 EN 1992-1-1:2004(E).
2. ACI 318-19.
3. ACI 318-19; 24.2.4.1.3.
4. Where there are notable differences between the bonded and unbonded designs, the difference is mentioned.
5. Long-term stress in prestressing steel after all friction and time-dependent stress losses.
6. ACI 318-19 Table 20.5.1.3.2.
7. ACI 318-19 Table 20.5.1.3.2.
8. EC2 EN1992-1-1;2004(E), Table 7.1N.
9. Hypothetical tensile stresses are explained in Chapter 4.
10. EC2 EN1992-1-1;2004(E), Table 7.1N.
11. For additional information refer to Chapter 4.
12. EC2 EN1992-1-1;2004(E), Section 7.2.
13. ACI 318-19 8.7.5.6.1.
14. ACI318-19 Section 8.6.2.1.
15. The assumption includes minor approximation.
16. CGS = Center of Gravity of Strand (Steel), which is distance from the closest extreme fiber to center of tension of prestressing.
17. RISA-ADAPT-Builder; www.RISA.com.
18. ACI 318-19 Section 6.4.3.2.

19. Midspan moments are not necessarily the largest of the respective span moments. But, for hand calculations midspan moments are commonly used.
20. EC2 EN1992-1-1;2004(E), Table 7.1N.
21. EC2 EN1992-1-1;2004(E), Section 7.1.
22. Find EC2 reference for relaxing deflection.
23. See Chapter 4 for EC2 derivation of hypothetical extreme fiber stresses.
24. EC2 EN1992-1-1:2004(E), Section 9.2.1.1.
25. EC2 EN1992-1-1:2004(E), Sections 9.2.1.1 and 9.3.1.1.
26. See Chapter 4 for EC2 serviceability flow chart.
27. See Chapter 4 for EC2 serviceability flow chart.
28. Refer to Chapter 4 for EC2 derivation of hypothetical stresses.
29. EC2 EN1992-1-1;:2004(E), 7.4.1(4).
30. EC2 EN1992-1-1;:2004(E), 7.4.1(5).
31. EC2 see serviceability flow chart in Chapter 4.
32. ACI 318-14 24.2.2.
33. ACI 318-19 24.2.4.1.3.
34. ACI 318-19 24.2.4.1.3.
35. ACI 318-19 8.3.4.1.
36. ACI 318 Table 8.6.2.3.
37. ACI 318 Table 8.6.2.3.
38. For skeletal members, combined with other simplifying assumptions, the hyperstatic actions can be calculated using a simpler relationship. The simple relationship is often used for hand calculation. Herein, the general method is used, since it applies to all conditions, in particular two- and three-dimensional structures, such as floor slabs and multi-level structures. For description of simple procedure see Chapter 6.
39. Chapter 6 includes an example of simple moment capacity calculation.
40. ACI 318-19 Table 5.3.1 and Section 5.3.11.
41. CGS Center of Gravity of Strand (Steel); center of tension.
42. Chapter 6 includes an example.
43. EC2 EN 1992-1-1:2004, Section 9.2.1-1(4).
44. EC2 EN1992-1-1;2004(E), Table 3.1.
45. For additional detail on this topic see Chapter 4 cracking moment.
46. ACI 318-19 Section 8.6.2.2.
47. EC2 EN1992-1-1;:2004(E), Section 6.4.3(6), Note.
48. EN 1992-1:2004, Section 6.4.5(3).
49. EC2 EN1992-1-1:2004(E) 6.4.4(1).
50. EC2 EN1992-1-1:2004(E) 6.4.4(1).
51. ACI 318-19 8.4.4 and 22.6.4.
52. Concrete International, Nov. 2005, P76.
53. ACI 318-19 8.4.4.2.2.
54. ACI 318-14 22.6.5.5.
55. ACI 318-14 8.4.4.2.2.
56. EC2 Table 7.2N or 7.3N; ACI 318-19 24.5.3.1 and 24.5.3.2.
57. ACI 318-11 Section 18.4.
58. The value specified is on the high side. Most hardware is designed to be stressed at 15 MPa concrete cylinder strength or less.
59. EC2 EN1992-1-1;:2004(E) 5.10.2.2 (5).

60. ACI 318-19 24.5.3.1 and 24.5.3.2.
61. ACI 318-19 Table 24.5.3.2.
62. ACI 318-19 Table 24.5.3.1.
63. ACI 318-19 8.4.2.2.3.
64. ACI 318-19 8.4.2.2.3.
65. ACI 318-19 8.7.5.6.1.

REFERENCES

Aalami, B. O. (2018), "The Challenge of the End Span," *Concrete International*, www.concrete.org, October 2018, pp. 52–54.

Aalami, B. O. (2016), "Cracking Moment and Safety of Post-Tensioned Members," *Structure Magazine*, www.structurema21.org, October 2016, pp. 34–35.

Aalami, B. O. (2014), *Post-Tensioned Buildings; Design and Construction*, Rewood City, CA, www.PTstructures.com, 450 pp.

ACI-318 (2019), *Building Code Requirements for Structural Concrete and Commentary*, American Concrete Institute, Farmington, MI, www.concrete.org, 624 pp.

European Code EC2 (2004), *Eurocode 2: Design of Concrete Structures – Part 1-1 General Rules and Rules for Buildings*, European Standard EN 1992-1-1:2004, CEN Brussels.

IBC (2021), *International Building Code*, International Code Council, iccsafe.org.

Chapter 6

Design of a post-tensioned beam frame

6.1 GEOMETRY AND STRUCTURAL SYSTEM

Figure 6.1 shows the typical framing of a beam and one-way slab parking structure. The structure features a central driveway with parking stalls on each side and two-way traffic in the middle.

One-way slabs span between the adjacent beams in the transvere direction. Figure 6.2 shows the elevation and cross-sectional geometry of the construction selected for the numerical example that follows. The representative structural system for design consists of a beam and its effective flange in the longitudinal direction.

In the longitudinal direction post-tensioning is limited to the beam stem. Transverse to the beams, the slab tendons are parallel and normal to the beam.

In the example the columns are assumed to extend below the parking deck only. At the foundation the columns are assumed on rollers. This reduces the restraint of the supports to shortening of the frame.

The design presented follows EC2[1] recommendations. At selected locations reference to ACI 318 points to the difference between the two codes in treatment of the subject. It is carried out for both bonded and unbonded post-tensioning systems.

6.1.1 Effective flange width

For bending effects of the applied load, an effective flange width is selected. The effective width and its associated beam stem are used to calculate the bending stresses in the structure.

For axial stresses the entire tributary of the flange is effective. With distance away from the tendon anchors at the beam ends, the precompression disperses through the entire cross-sectional area. Figure 6.3 identifies the different sections used for bending and axial effects.

Figure 6.4 defines the effective width for bending of flanged beams using EC2.[2] Figure 6.5 illustrates the parameters for the determination of the effective width.

DOI: 10.1201/9781003310297-6

134 Post-tensioning in building construction

Figure 6.1 Framing example of beam and slab parking structure.

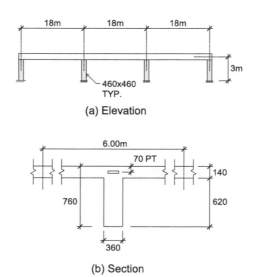

**Typical Frame Elevation and Section
(mm U.N.O.)**

Figure 6.2 View of beam and slab construction showing the beam's tributary.

Using EC2's recommendation, the following determines the applicable effective width:

$$b_{eff} = \sum b_{eff,i} + b_w < b \qquad (6.1)$$

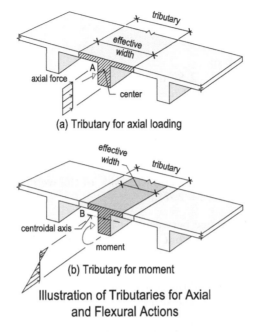

Figure 6.3 Illustration of tributaries for axial and bending effects of flanged beams.

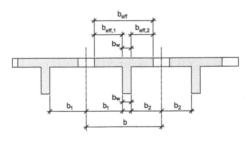

Figure 6.4 Illustration of effective width in bending.

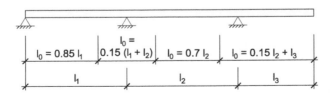

Figure 6.5 Illustration of effective width parameters for bending effects.

$$b_{ef,i} = 0.2b_i + 0.1l_0 \leq 0.2l_0 \tag{6.2}$$

where

$l_0 = 18 \times 0.7 = 12.6$ m
$b_i = 0.5(6.00 - 0.36) = 2.82$ m
$b_{eff,i} = 0.2 \times 2.82 + 0.1 \times 12.6 = 1.824$ m
$b_{eff} = 2\, b_{eff,i} + b_w = 2 \times 1.824 + 0.36 = 4.008$ m

Using ACI's recommendation,[3] the applicable value for the effective width for the current condition is:

b_{eff} = 8 times slab thickness on each side of the web plus web thickness.
$b_{eff} = 8 \times 2 \times 0.14 + 0.36 = 2.6$ m

There is a large difference between the EC2 and ACI recommended value. More importantly, neither code explicitly differentiates between the effective width for bending and axial effects.

In light of the above, the following is assumed for the design that follows. The sequence of design and the considerations remain independent from the assumption of effective width values:

Effective width for bending effects: $b_{eff,bending} = 2.82$ m
Effective width for axial effects: $b_{eff,axial} = 6.00$ m

6.1.2 Section properties

The section properties for the axial effects are the same for all spans. These are based on full cross-sectional geometry of a typical beam stem and its tributary flange.

For bending effects, the sectional properties are based on the reduced geometry defined by the effective width. The section properties are summarized in Table 6.1.

6.2 MATERIAL PROPERTIES

6.2.1 Concrete

Cylinder strength $f_{ck} = 30$ MPa
Weight = 2,400 kg/m³
Modulus of elasticity = $22 \times 10^3\ [(f_{ck}+8)/10]^{0.34} = 32{,}836$ MPa
Creep coefficient = 2
Material factor, $\gamma_c = 1.50$
Concrete strength at tendon stressing $f_{ci} = 20$ MPa

Table 6.1 Section properties

Item	Unit	Typical Span Axial effects	Bending effects
Area	mm²	1.0632×10⁶	-------
I	mm⁴	-----	2.838×10¹⁰
Y_t	mm	-----	207
Y_b	mm	-----	553
S_{top}	mm³	-------	1.371×10⁸
S_{bot}	mm³	-------	5.132×10⁷

I = second moment of area (moment of inertia);
Y_t = distance of centroid to top fiber of section;
Y_b = distance of centroid to bottom fiber of section;
S_{top} = section modulus for top fiber (I/Y_t);
S_{bot} = section modulus for bottom fiber (I/Y_b).

6.2.2 Non-prestressed reinforcement

f_{yk} = 460 MPa
Modulus of elasticity = 200,000 MPa
Material factor, γ_c = 1.15
Cover to reinforcement = 25 mm

6.2.3 Prestressing

Material – Low relaxation, seven-wire strand
Strand nominal diameter = 13 mm
Strand area = 99 mm²
Modulus of elasticity = 193,000 MPa
Ultimate strength of strand f_{pk} = 1,860 MPa
Material factor, γ_p = 1.15 for
Cover to tendon CGS = 70 mm

Long-term effective stress after all losses:

Unbonded system = 1,200 MPa
Bonded system = 1,100 MPa

6.2.4 Cover to reinforcement

Cover to rebar = 25 mm
   Minimum distance to center of prestressing tendons (CGS) = 70 mm. This facilitates the positioning of tendons within the cage of the beam.

6.3 LOADS

6.3.1 Dead load

Self-weight: based on full volume of the tributary
Volume = (6.00×0.14 + 0.36×0.62) ×24,000/1,000[5] = 25.52 kN/m

6.3.2 Live load

Assume live load uniformly distributed over the entire parking deck [Inst Struct Engrs, 2002]
LL = 2.5 kN/m²

Live load for the beam tributary is:
LL = 2.5×6.00 = 15 kN/m

There will be no reduction of live load intensity.
Max LL/DL ratio = 15/25.52 = 0.59

Since ratio of maximum live to dead load is less than 0.75, skipping of live load to maximize force demand is not required.[6]

6.4 DESIGN PARAMETERS

6.4.1 Applicable code

The design is based on EC2(EN 1992-1-1, 2004). Where EC2 is mute, depending on the condition the Institution of Structural Engineering Associations Publication [Inst Structural Engineers, 2002] or ACI 318 [ACI (318) 19, 2019] is used.

6.4.2 Allowable stresses

EC2 does not specify 'limiting' allowable stresses in the strict sense of the word. There are stress thresholds that trigger crack control. The values specified are the same for both bonded and unbonded systems.

(i) For 'frequent' (total) load condition

Concrete:

Compression = $0.60 f_{ck}$ = 0.6×30 = 18.00 MPa
Tension (concrete) = $f_{ct,eff} = f_{ctm}$[7] = $0.30 \, f_{ck}^{(2/3)}$[8] = $0.30 \times 30^{(2/3)}$ = 2.90 MPa

Reinforcement:

    Tension (mild steel) = $0.80 f_{yk}$ = 0.8×460 = 368 MPa
    Tension (prestressing steel) = $0.75 f_{pk}$ = 0.75×1,860 = 1,395 MPa

(ii) For 'quasi-permanent' (sustained) load condition

    Compression = $0.45 f_{ck}$ = 0.45×30 = 13.5 MPa
    Tension (concrete) = 2.90 MPa; same as frequent load combination

Unlike ACI 318, EC2 provisions permit[9] overriding the hypothetical tension stress thresholds in concrete, provided cracking is controlled so as not to exceed the selected 'design crack width.' Chapter 4 on building codes offers additional information.

(iii) For 'initial' load condition

    Tension (unbonded and bonded) = $f_{ct,eff}$ = f_{ctm} = $0.30 \, f_{ci}^{(2/3)}$ [10]
    $f_{ctm} = 0.30 \times 20^{(2/3)}$ = 2.21 MPa
    Compression[11] = $0.60 \, f_{ci}$ = 0.6 × 20.00 = 12.00 MPa

In summary, for both unbonded and bonded post-tensioning systems, EC2[12] recommends $f_{ct,eff}$ as the threshold for hypothetical tensile stresses, before it becomes necessary to provide reinforcement for crack control. For stresses below this threshold, the minimum reinforcement provisions of EC2 suffice.

6.4.3 Cracking limitation

The allowable crack width depends on whether the post-tensioning system used is 'bonded' or 'unbonded' and the load combination being considered.[13] The load combinations and the respective allowable crack widths are:

    Frequent load condition: for prestressed members with bonded tendons allowable crack width is 0.2 mm.

    Quasi-permanent load condition: for prestressed members with unbonded tendons the allowable crack width is 0.3 mm.

6.4.4 Allowable deflection

For parking decks the allowable deflection is not critical as long as adequate drainage for surface water is accounted for. The Institution of Structural Engineers [Institution of Structural Engineers, 2002] recommends long-term drainage slope not less than 1/64.

6.5 ACTIONS FROM DEAD AND LIVE LOADS

The structural system of the frame and its load are shown in Figure 6.6. The members are assumed prismatic with uniform cross section. Centerline to centerline distances is used for span lengths.

The actions are calculated using a generic plane frame analysis program. Figure 6.7 shows the distribution of moments from the dead and live loads. Moment values shown are at the face of support and midspan.

The maximum design moments are not generally at midspan. But, for hand calculation, the midspan location is selected. The approximation is acceptable when spans and loads are essentially uniform.

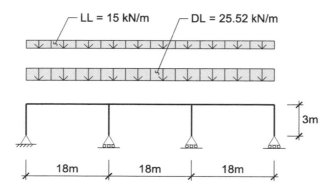

Figure 6.6 Structural system and loads.

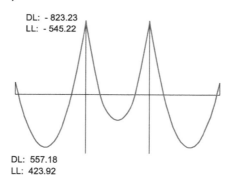

Figure 6.7 Dead and live load moments at face of support and midspan.

6.6 POST-TENSIONING

6.6.1 Selection of post-tensioning tendon force and profile

Unlike conventionally reinforced members, where given geometry, boundary conditions, material properties and loads result in a unique design, for post-tensioned members two additional input assumptions are required to conclude a design. This is because the number of input parameters necessary to complete a design is short of the available equations by two. The assumed input parameters are generally based on the experience of similar successful construction.

For the two entry parameters the common practice is:

(i) Assume an average precompression for post-tensioning.
(ii) Target to balance a given fraction of the structure's dead load by post-tensioning.

From a construction standpoint, the force selected will be fine-tuned to divide into a round number of prestressing strands, and where bonded tendons are used, to the efficient use of number of ducts.

Neither EC2 nor ACI 318 require minimum precompression for design of post-tensioned beams. For beams of the current dimensions and application 1.75 MPa precompression works well as one of the entry assumptions. The common range of precompression for the typical parking structure beams is between 1.50 and 2.00 MPa. The suggested precompression will be adjusted to a round number of strands in the beam stem. The selection is also governed by the relative unit cost of post-tensioning and non-prestressed reinforcement. In the following the lower bound is selected, assuming that unit weight of post-tensioning costs much more than non-prestressed reinforcement.

Assume 1.50 MPa average precompression.

Total force per beam stem = beam tributary area 1.50 MPa
Total force = $1.0632 \times 1.50 \times 10^6/10^3$ = 1,594.8 kN

Long-term effective stress in strand = 1,100 MPa

Effective force per strand = $99 \times 1,100/1,000$ = 108.90 kN
Number of strands = 1,594.8/108.90 = 14.6 strands

Select 15 strands. This offers the option of using three bonded ducts, each containing five strands.

The force provided by 15 strands is:

PT = 15×108.90 = 1,633.50 kN
Average precompression = (PT force/tributary area) = $1,633.50 \times 1,000/1.0632 \times 10^6$ = 1.54 MPa OK

6.6.2 Post-tensioning actions

A. **Tendon profile**: Figure 6.8 shows the centroid (CGS) of the tendon path selected. At each end, tendons are anchored at the centroid of the tributary area.

B. **Post-tensioning actions**: Actions from post-tensioning are determined using commonly available plane frame software for strip design of post-tensioned members.

First, as outlined in Chapter 3, the balanced loads from the post-tensioned tendons selected are calculated. The balanced loads are applied to the frame of the structure to calculate the actions from post-tensioning. Figure 6.9 shows the moments from post-tensioning tendons (kNm). As expected, these counteract the moments from gravity loads shown in Figure 6.7.

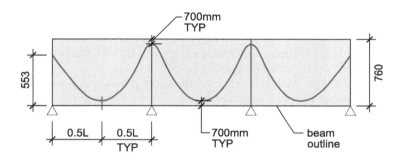

Figure 6.8 Tendon profile.

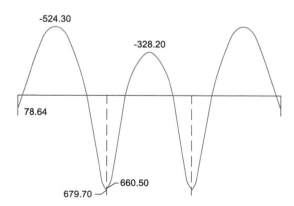

Figure 6.9 Distribution of moments from post-tensioning. Values shown are at the face of support or midspan (kNm).

6.7 CODE CHECK FOR SERVICEABILITY LIMIT STATE (SLS)

6.7.1 Deflection check

For parking structures, the focus of deflection check is its impact on the drainage of surface water. Adequate drainage requirements generally override the deflection from other loads defined earlier. A minimum slope of 1/64 is acceptable.[14]

6.7.2 Stress check/crack control

For hand calculation, the critical locations for stress check are based on engineering judgment using the moment diagrams as guide. The selected locations may or may not coincide with the locations of maximum stress. This introduces a certain degree of approximation in design. Computer solutions generally calculate stresses at multiple locations along a span, thus providing closer values in determination of maximum combined moment and stresses.

From inspection of the moment diagrams, stress checks at the bottom fiber at the middle of the first span, and top fiber at the face of the first interior support in span 1 are selected for evaluation.

The extreme fiber stress is given by the following combination to be duly adjusted with the code-specified factors:

$$\sigma = (M_{DL} + M_{LL} + M_{PT})/S + P/A \tag{6.3}$$

$$S = I/Y_c \tag{6.4}$$

where

A is the area of the entire beam tributary;
I is the second moment of area of the portion of the cross-section that is defined by the effective width for bending;
M_{DL}, M_{LL} and M_{PT} are the moments across the entire tributary of the design strip;
P is the axial force on the entire tributary of the beam;
S is the section modulus of the cross section reduced through effective width defined for bending action; and
Y_c is the distance of the centroid of the reduced section (defined for bending) to the farthest tension fiber of the section.

Figure 6.10 shows the cross-section of the member and the position of its centroid.

From the figure, the parameters for the code-required stress check are:

$A = 1.0632 \times 10^6$ mm^2
$I = 2.838 \times 10^{10}$ mm^4

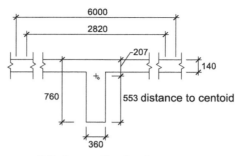

Figure 6.10 Cross-section of the typical design element.

$S_B = 5.132 \times 10^7$ mm³
$S_T = 1.371 \times 10^8$ mm³
$Y_B = 553$ mm
$Y_T = 207$ mm

Suffixes B and T refer to the farthest bottom and top fibers of the section to the centroidal axis.
Average precompression:

$P/A = 1.54$ MPa

A. Based on EC2: Stress checks are carried out for frequent and quasi-permanent load combinations. The outcome will determine whether or not crack width must be controlled.

The calculations below follow the flow chart given in Chapter 4 for serviceability check of post-tensioned members.

Consider two critical locations, namely the bottom fiber at the center of span 1 and the top fiber at the face of support at the right of span 1.

Stress thresholds:
Frequent load combination

Compression = $0.60 \times 30 = -18$ MPa
Tension (concrete) = $f_{ct,eff} = f_{ctm} = 2.90$ MPa

Quasi-permanent load combination

Compression = $0.45 \times 30 = -13.5$ MPa
Tension (concrete) = $f_{ct,eff} = f_{ctm} = 2.90$ MPa

Span 1: midpoint
Frequent load combination

$$\sigma = (M_{DL} + 0.5\ M_{LL} + M_{PT})/S + P/A$$

$M_D + 0.5\ M_L + M_{PT} = (557.18 + 0.5 \times 423.92 - 524.30) = 244.84$ kNm
Top fiber:
$\sigma = -244.84 \times 10^6 / 1.371 \times 10^8 - 1.54 = -3.32$ MPa Compression < -18 MPa OK

Bottom fiber:
$\sigma = 244.84 \times 10^6 / 5.132 \times 10^7 - 1.54 = 3.23$ MPa Tension > 2.90 MPa NG
Hence crack width check and control required.[15]
Crack control detailing for this location is carried out at the end of stress checks.

Quasi-permanent load condition

$$\sigma = (M_{DL} + 0.3\ M_{LL} + M_{PT})/S + P/A \qquad (6.5)$$

$M_D + 0.3M_L + M_{PT} = (557.18 + 0.3 \times 423.92 - 524.30) = 160.06$ kNm
Top fiber:
$\sigma = -160.06 \times 10^6 / 1.371 \times 10^8 - 1.54 = -2.71$ MPa Compression < -13.50 MPa OK
Bottom fiber:
$\sigma = 106.06 \times 10^6 / 5.132 \times 10^7 - 1.54 = 0.53$ MPa Tension < 2.90 MPa OK

Span 1: face of second support
Frequent load combination
$$\sigma = (M_{DL} + 0.5\ M_{LL} + M_{PT})/S + P/A \qquad (6.6)$$
$M_D + 0.5\ M_L + M_{PT} = (-823.23 - 0.5 \times 545.22 + 679.70) = -416.14$ kNm
Top fiber:
$\sigma = 416.14 \times 10^6 / 1.371 \times 10^8 - 1.54 = 1.49$ MPa Tension < 2.90 MPa OK
Bottom fiber:
$\sigma = -416.14 \times 10^6 / 5.132 \times 10^7 - 1.54 = -9.64$ MPa Compression < -18 MPa OK

Quasi-permanent load condition

$$\sigma = (M_{DL} + 0.3\ M_{LL} + M_{PT})/S + P/A$$

$M_D + 0.3M_L + M_{PT} = (-823.23 - 0.3 \times 545.22 + 679.7) = -307.10$ kNm
Top fiber:
$\sigma = 307.10 \times 10^6 / 1.371 \times 10^8 - 1.54 = 0.70$ MPa Tension < 2.90 MPa OK
Bottom fiber:
$\sigma = -307.10 \times 10^6 / 5.132 \times 10^7 - 1.54 = -7.52$ MPa Compression < -13.50 MPa OK

Span 1: crack control detailing

EC2 offers two options for crack control, when the hypothetical extreme fiber tension stress exceeds the code's threshold (Chapter 4). Either calculate the probable crack width to ensure that it is less than the recommended value, or detail the location for crack control. The latter option is selected.

The required bonded reinforcement for crack control is given by the following relationship recommended in TR43:[16]

$$A_{scrack} = \frac{N_c}{(5/8)f_{yk}} \qquad (6.7)$$

where N_c is the tensile force in the tension zone of the member.

Top fiber stress = 3.32 MPa compression
Bottom fiber stress = 3.23 MPa tension
Member depth = 760 mm
From Figure 4.7 of Chapter 4 depth of the tension zone is given:
Depth of tension zone = [3.23/ (3.23 + 3.32)]760 = 375 mm
N_c = 0.5(360 × 375) ×3.23/1,000 = 218.025 kN
A_{scrack} = 218.025×1,000/[(5/8) ×460] = 758.35 mm²
Cross-sectional area of 15 strands bonded reinforcement available = 1,485 >758.35 mm² OK
Crack control reinforcement not required.

B. Based on ACI

The requirement of crack control depends on the magnitude of the extreme fiber hypothetical tension stress (see Chapter 4 for details). The first threshold for considering cracking is the 'transition' stress for initiation of cracking.[17] It is $0.625\sqrt{f_c'}$:

Transition threshold = $0.625\sqrt{30}$ = 3.42 MPa

The next threshold, when reduction in stiffness from cracking in estimate of member deflection need be accounted for, is $\sqrt{f_c'}$ = 5.55 MPa

ACI 318 does not specify the load combination for crack control. Commonly, sustained load combination is used.

Consider the bottom fiber at midpoint of span 1, using expression 6.3

$$\sigma = (M_{DL} + M_{LL} + M_{PT})/S + P/A$$

$$M_D + M_L + M_{PT} = (557.18 + 423.92 - 524.30) = 456.80 \text{ kNm}$$

Top fiber:
σ = - 456.80 × 10⁶/1.371×10⁸ – 1.54 = - 4.87 MPa compression

Bottom fiber:
σ = 456.80 × 10⁶/5.132×10⁷ – 1.54 = 7.36 MPa tension

Computed hypothetical extreme fiber at midspan is 7.36 >5.55 MPa

The computed stress exceeds the cracking threshold. The section is considered cracked. ACI 318 does not require the addition of non-prestressed reinforcement. The code requires that the loss in stiffness due to cracking be accounted for in the estimate of member deflection.

6.8 CODE CHECK FOR STRENGTH (ULS)

The following is according to EC2 recommendations. ACI 318 follows essentially the same steps, except for somewhat different load combination factors (Chapter 4).

6.8.1 EC2 load combination

$$U = 1.35DL + 1.5LL + 1.0HYP \tag{6.8}$$

where HYP is the hyperstatic forces from post-tensioning.

6.8.2 Calculation of hyperstatic actions

In the current scenario, the member is prismatic. Tendons are continuous from one end to the other. It is expeditious to use the indirect method to compute the hyperstatic moments from post-tensioning.

Chapter 3 details the calculation of hyperstatic moments. Using the indirect method, the relationship is:

$$M_{hyp} = M_{pt} - Pe \tag{6.9}$$

where

M_{hyp} = hyperstatic moment;
M_{pt} = post-tensioning moment;
P = post-tensioning force; and
e = eccentricity of post-tensioning force.

The calculations are carried out at the left face of the second support where moments are highest.

M_{pt} = 679.70 kNm
P = 1,633.5 kN

Centroid of tendon 70 mm below the top surface. From Figure 6.10 tendon eccentricity is:

e = 207 − 70 = 137 mm
M_{hyp} = 679.70 − 1,633.50×0.137 = 455.91 kNm

This is at the center of support. At the face of support, the value will be slightly less. For hand calculation the same value is used. Figure 6.11 shows the distribution of hyperstatic moments.

The distribution of the hyperstatic moments is a straight line between the center of supports.

6.8.3 Calculation of design moments

From expression 6.6 the design moment for the face of support on the right of span 1 is:

$M_u = -1.35 \times 823.23 - 1.5 \times 545.22 + 455.98 = -1,473.21$ kNm

The design moment at other locations is calculated similarly.

6.8.4 Strength design for bending and ductility

A. Check for strength

The strength design for bending consists of two provisions.

(i) The design capacity shall not be less than the demand. A combination of prestressing and non-prestressed steel provides the design capacity.
(ii) The ductility provision of the section in bending shall not be less than the limit set in the building code.

The ductility requirement is met if section failure in bending is initiated by yielding of its reinforcement, as opposed to crushing of concrete. This is

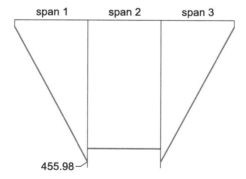

Figure 6.11 Distribution of hyperstatic moments.

achieved through the limitation imposed on the depth of the compression zone.[18]

For expeditious hand calculation, the flexural capacity of a post-tensioned member in common building structures can be approximated by assuming a conservative maximum stress for prestressing tendons.

Unlike conventionally reinforced concrete, where at each section along a member non-prestressed reinforcement must be available to resist the design moment, in prestressed members this is not the case. Prestressed members possess base capacity provided by tendons along the entire length of the prestressing steel. Non-prestressed reinforcement is needed at sections, where the moment demand exceeds the base capacity of the section. For this reason, in hand calculation, it is expeditious to first calculate the base capacity. Add reinforcement, if required.

Safety check requires two considerations. These are: safety check for factored demand moments; and safety check for cracking moment. Unlike ACI 318, EC2 does not require safety check for cracking moment for beams reinforced with bonded tendons. Hence, in this case only safety check for factored demand moment is considered.

The calculated design moment from the previous section is -1,473.21 kNm

Leading section dimensions are:

b = 360 mm
h = 760 mm

Contribution of tension from prestressing

$f_{pd} = f_{pk}/1.15 = 0.9 \times 1,860/1.15 = 1,455.65$ MPa
$T_p = 15 \times 99 \times 1,455.65/1,000 = 2,161.64$ kN

Contribution of tension from non-prestressed reinforcement is zero at this stage, since none is specified.

Allowable tension stress for non-prestressed reinforcement is:

$f_{yd} = f_{yk}/1.15 = 0.87 f_{yk}$
$f_{yd} = 460/1.15 = 400$ MPa

Design force for prestressing strands is:
$T_p = 15 \times 99 \times 1,455.65/1,000 = 2,161.64$ kN

Total tensile force (T) at ultimate limit state is the sum of tension from prestressing (T_p) and tension from rebar (T_s).

$T = T_p + T_s$
$T = 2,161.64 + 0 = 2,161.64$ kN

Design value for stress for the compression block is:

$f_{cd} = 0.85 \, f_{ck}/1.5 = 0.567 \, f_{ck}$
$f_{cd} = 0.567 \times 30 = 17.01$ MPa

Depth of compression zone 'a' is:
$a = T/(\eta f_{cd} \, b) = (2{,}161.64 \times 1{,}000)/(1 \times 17.01 \times 360) = 353.00$ mm

The expression for moment capacity M_{Rd} is:

$$M_{Rd} = (T_p z_p + T_s z_s) \qquad (6.10)$$

where
$\quad z_p$ and z_s are moment arms for prestressing steel and rebar respectively.
$\quad M_{Rd} = [2{,}161.64 \, (760 - 70 - 0.5 \times 353.00) + 0]/1{,}000 = 1{,}110.00$ kNm $<$ demand moment 1,473.21 kNm
The shortfall in moment capacity must be provided by non-prestressed reinforcement.
$\quad$Shortfall in moment capacity
$\quad M_{rebar} = 1{,}473.21 - 1{,}110.00 = 363.21$ kNm

Rebar cover 25 mm
$\quad$Use 26 mm rebar
$\quad$Moment arm from previous lines is approximately:
$\quad Z_{rebar} = 760 - 25 - 0.5 \times 26 - 0.5 \times 353 = 545.50$ mm

Area of non-prestressed reinforcement required
$\quad A_s = M_{rebar}/(f_{yd} \times Z_{rebar}) = 363.21 \times 1{,}000^6/(400 \times 545.50) = 1{,}665$ mm^2

Required number of bars:
$\quad$Select 26 mm bars, each 531 mm^2

$\quad$1,665/531 = 3.13 bars; select 4 bars
$\quad A_s$, provided = 4×531 = 2,124 mm^2

Total area of reinforcement = strands + rebar = 1,485 + 2,124 = 3,609 mm^2

$\quad$B. Check for ductility

To meet the ductility requirement of EC2 the depth of the compression zone (x) should be less than the EC2-specified threshold.
$\quad$Depth of compression zone a = 353.00 mm

$\quad$For f_{ck} = $<$ 50 MPa $x = 0.43h$

$x/h = 353.00 / 760 = 0.46 > 0.43$ NG

It is recommended to increase the width of the section from 360 mm to 400 mm, in order to meet the ductility requirement of EC2.

6.8.5 One-way shear design

The shear design is based on EC2 recommendations.[19] The load combination for shear design is the same as for bending effects. Figure 6.12 shows the distribution of factored shear. The values shown in the figure refer to the face of supports.

Consider span 1. The governing location is the face of the interior support. The shear values for span 1 are:

Right of first support 471.01 kN
Left of second support 578.93 kN

Figure 6.13 shows the distribution of factored shear for the first span.
Support width = 460 mm

Point of zero shear from the face of second support is:

$578.93 \times 17.54 / (578.93 + 471.01) = 9.67$ m

Depth of member (d) for shear design at the right support

$d = 760 - 70 = 690$ mm

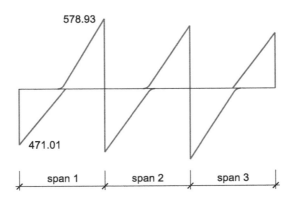

Distribution of Factored Shear (kN)

Figure 6.12 Cross-sectional geometry of design element.

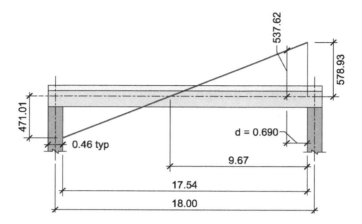

Figure 6.13 Distribution of factored shear (kN).

Distance of first design section from face of column = d
Design demand at first design location

$$V_{ED} = 578.93 \times (9.67 - 0.690)/9.67 = 537.62 \text{ kN}$$

Shear resistance provided by the concrete section only is $V_{Rd,c}$

$$V_{Rd,c}{}^{20} = [\ C_{Rd,c}\ k\ (100\ \rho_1\ f_{ck}\ )^{1/3} + k_1\ \sigma_{cp}\ ]b_w d \qquad (6.11)$$

but not less than $(v_{min} + k_1 \sigma_{cp}) b_w d$
where
f_{ck} = 30 MPa
$k = 1 + (200/d)^{1/2} = 1 + (200/690)^{1/2} = 1.54 < 2.0$ OK
Area of longitudinal reinforcement from bending design cosisting of tendons and added rebar is: 2,124 mm²

Area of post-tensioning tendons = 1,485 mm²
A_{sl} = 2,124 + 1,485 = 3,609 mm²
$\rho_1 = A_{sl} / (b_w\ d) = 3,609/(360 \times 690) = 0.0145$
σ_{cp} = average precompression

$\sigma_{cp} < 0.2 f_{cd} = 0.2 \times (30/1.5) = 4$ MPa
Design precompression σ_{cp} = 1.74 < 4 MPa OK

$C_{Rd,c} = 0.18/\gamma_c = 0.18/1.50 = 0.12$
$k_1 = 0.15$
$V_{Rd,c} = [\ 0.12 \times 1.54\ (100 \times 0.0137 \times 30)^{1/3} + 0.15 \times 1.74\ ]\ 360 \times 690/1,000 =$ 223.25 kN

<541.84 kN hence shear reinforcement required.

$$v_{min} = 0.035 \, k^{3/2} \, f_{ck}^{1/2} = 0.035 \times 1.54^{3/2} \times 30^{1/2} = 0.366 \text{ MPa}$$

Based on v_{min}
V_{min} = 0.366×360× 690/1,000 = 90.91 < 541.84 kN hence shear reinforcement required

$$V_{Rd,min} = (v_{min} + 0.15 \, \sigma_{cp}) b_u d$$

$V_{Rd,min}$ = (0.366+0.15×1.74)360×690 /1,000 = 155.75 <541.84 kN hence shear reinforcement required.

Assume 13 mm stirrups with two legs:

f_{yk} = 460 MPa
A_{sw} = 2 × 132.7 mm² = 265.5 mm²

The spacing,[21] s, between the stirrups is given by:

$$s = (A_{sw} / V_{Rd,s}) z \, f_{ywd} \cot\theta$$

Assume $\theta = 40^0$, $\cot \theta = 1.20$

$V_{Rd,s} = V_{ED} - V_{Rd,c}$ = 537.62– 223.25 = 314.37 kN
z = 0.9 d = 0.9 × 690 = 621 mm
s = (265.5/[314.37 × 1,000]) × 621 (460 / 1.15) × 1.20 = 251.74 mm

Select spacing 2-legged 13 mm stirrups at 200 mm over the entire length of the beam.

NOTES

1. EN 1992-1-1:2004(E)
2. EC2 5.3.2.1
3. ACI 318-16 6.3.2.1
4. EN 1992-1-1:2004(E), Table 3.1
5. 1 cubic m of concrete assumed 2,400 kN
6. ACI 318-19 6.4.3.2
7. EN 1992-1-1:2004(E), Section 7.3.2(4)
8. EN 1992-1-1:2004(E), Table 3.1
9. EN 1992-1-1:2004(E), Section 7.3.2(4)
10. EN 1992-1-1:2004(E), Table 3.1
11. EN 1992-1-1:2004(E), Section 5.10.2.2(5)
12. EN 1992-1-1:2004(E), Section 7.3.2(4)
13. EN1992-1-1:2004(E), Table 7.1N

14. Institution of Structural Engineers given in the list of references
15. EN 1992-1-1:2004(E), Section 7.3.4
16. TR43 given in list of references
17. ACI 318-19; 24.5.2.1
18. ACI 318-11; 9.3.2.
19. EN 1992-1-1:2004(E), Section 6.2
20. EN 1992-1-1:2004 (E), Section 6.2.2
21. EN 1992-1-1:2004 (E), Exp: 6.12

REFERENCES

ACI-318 (2019), *Building Code Requirements for Structural Concrete and Commentary*, American Concrete Institute, Farmington, MI, www.concrete.org, 624 pp.

European Code EC2 (2004), *Eurocode 2: Design of Concrete Structures – Part 1-1 General Rules and Rules for Buildings*, European Standard EN 1992-1-1:2004, CEN Brussels.

The Institution of Structural Engineers (2002), *Design Recommendations for Multi-Story and Underground Car Parks*, The Institution of Structural Engineers, London, 3rd ed., 86 pp.

Chapter 7

Member shortening; precompression: member strength

7.1 POST-TENSIONING; SHORTENING; PRECOMPRESSION

In post-tensioning, tendons are stressed and anchored after concrete has developed adequate strength. If at stressing the member is free to shorten under the jacking force, the tension in the post-tensioned tendons results in an equal compression in the member. The compression shortens the member.

In most applications, the tendons are profiled between the stressing ends. In addition to precompression, profiled tendons exert lateral forces on the member – forces normal to the line of tendons. The lateral forces bend the member in addition to the shortening from the jacking force at the member ends.

The following explains that the contribution of post-tensioning to flexural strength of the member is closely related to the precompression in the member from post-tensioning. The phenomenon was well recognized in the early stages of application of post-tensioning [Leonhardt, F., 1964].

7.2 RELATIONSHIP BETWEEN SHORTENING AND PRECOMPRESSION

Figure 7.1 (a) shows the force-displacement relationship of the common compression test of concrete cylinders. Figure 7.1 (a) illustrates that the compression force (P) on the cylinder relates to the shortening (u). Similarly, Figure 7.1 (c) shows that the compression force F is the function of the member displacement u. There will be no compression force in the member if the member is not free to shorten under the applied axial force.

Post-tensioned members, such as floor slabs and beams, are generally supported on walls and columns. Depending on the connectivity of the post-tensioned member to other structural parts, and the details of the member support, the supports can restrain the free shortening of the member at jacking.

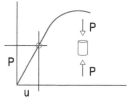

(a) Force-displacement relationship

(b) Member with no pre-compression

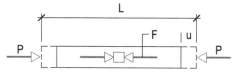

(c) Member with pre-compression

Pre-compression and Shortening

Figure 7.1 Precompression and shortening.

Only if the member is free to shorten will it receive the entire precompression from the stressed tendons.

Figure 7.2 illustrates a member that can shorten (part a) and one that cannot (part b). In part (b) there will be no compression in the member; shrinkage can lead to through cracks.

Restraint cracks leave the member with no precompression from post-tensioning, while the tendon within the member retains its full tension force. In most cases the cause of restraint crack is shrinkage of concrete.

Apart from possible aesthetic objections, the restraint cracks can cause leakage, and expose the reinforcement to corrosive elements.

More importantly, restraint cracks reduce the contribution of post-tensioning tendons to the strength capacity of the member – the focus of what follows next.

The extent of the restraint cracking in a post-tensioned member depends on a number of factors, including the stiffness of the supports. If the supports fully prevent shortening, the entire post-tensioning force is diverted to the supports, leaving the member with no precompression.

Restraint cracks are long; typically, they extend through the full length of the floor panel and beyond. These cracks occur at locations of the slab's axial weakness, such as where non-prestressed reinforcement is reduced or

Member shortening; precompression: member strength 157

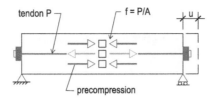

(a) Member free to shorten

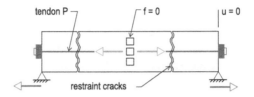

(b) Restrained member

Restrained and Unstrained Members

Figure 7.2 Restrained and unrestrained members; impact of support restraint.

Figure 7.3 Example of a restraint crack.

terminated, or where there is a reduction in the slab's cross-sectional area. Figure 7.3 is an example of restraint crack that extends through the depth of the slab.

Restraint cracks are most pronounced at the first floor above the foundation. This is primarily due to the fixity at the far end of the floor supports to the foundation. There is less cracking at higher floor levels.

Experienced design engineers are aware of the possibility of restraint cracking and its adverse consequence. They use a number of measures that allow the post-tensioned member to shorten, while minimizing the cracking and its effects in either the member or its supports [Aalami, B. O., 2017, 2018, 2021].

7.3 PRECOMPRESSION AND MEMBER STRENGTH

The following describes the mechanism of the loss in moment capacity of post-tensioned members from support restraint.

7.3.1 No support restraint

Figure 7.4 illustrates the mechanism by which post-tensioning tendons contribute to the strength of a member when there is no support restraint. This will be contrasted to the case in Figure 7.5, where the member is subject to support restraint.

The member shown in part (a) is on rollers. There are no horizontal forces at the supports. The entire post-tensioning force at the anchorage is transferred to the member in the form of compression.

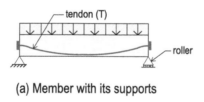

(a) Member with its supports

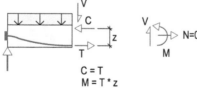

(i) Resistance at cut (ii) Force demand

(b) Free body diagram of member segment

Post-Tensioned Member with
No Support Restraint to
Shortening

Figure 7.4 Post-tensioned member with no support restraint to shortening.

Member shortening; precompression: member strength   159

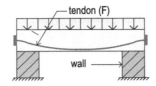

(a) Member with restraining walls

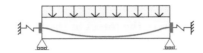

(b) Simplified model of member with axial restraint

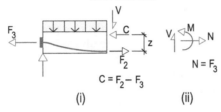

(c) Free body diagram of severed member

Post-Tensioned Member with Support Restraint

Figure 7.5 Post-tensioned member with support restraint.

Consider the safety requirements of the member at its midspan. The forces that act on the left side of the member are shown in part (b-i) of the figure. The forces shown on the severed left half of the member must satisfy static equilibrium.

The demand forces at the midspan cut are shown in part (b-ii). These are the moment (M), shear (V) and axial force (N).

The demand actions M, V, and N are in static equilibrium with the forces acting on the severed segment of the member on the left.

For the safety of the structure, the resistance that develops at the face of the cut, namely forces T, C, and V should not be less than the demand actions M, V, and N.

Since the member is assumed on rollers, the reaction at the support shown in Figure 7.4 (b) is limited to a vertical force. There is no horizontal restraint at the supports, so there is no horizontal force demand on the face of the cut segment. Hence, in Figure 7.4 (b-ii) the axial force on the demand side is zero.

$$N = 0 \qquad (7.1)$$

The forces developed at the face of the cut must balance the force demand for equilibrium of the segment, namely V, M, and N.

The resistance to the demand moment (M) at the face of the cut is developed by the tendon force (T) and the compression force (C) in concrete:

$$T = C \tag{7.2}$$

$$M = Tz \tag{7.3}$$

where z is the moment arm of the forces at the face of the cut. In the above expression 7.3, the *entire force in tendon* (T) is available to contribute to the moment resistance.

7.3.2 Finite support restraint

Unless the supporting columns and or walls of a post-tensioned member are extremely flexible, they restrain the free shortening of the member when the tendons are pulled. If restrained, the member does not receive the full amount of design precompression from the tendons.

Figure 7.5 shows full support restraint to member shortening. For ease of visualization, the member is modeled as shown in part (b). The springs attached at each end of the member represent the restraint of the supports to the shortening of the member. There will also be moment at the end of the member due to the shift of the restraining force (F_3) at the support from the support/member interface to the centroid of the member shown in part (b). This moment is not shown in the figure, since its presence does not impact the current discussion.

A portion of the post-tensioning force, marked F_3 in part (c) of the figure will be diverted to the supports. The value of the force F_3 depends on the stiffness of the supports. The remainder of the post-tensioning force results in precompression in the member.

Part (c) of the figure is the free body diagram of the left half of the member.

The demand actions at the face of the cut from the forces on the cut segment are again the moment M, shear V, and axial force N (part c-ii). In this case, however, the equilibrium of the forces in the horizontal direction gives:

$$N = F_3. \tag{7.4}$$

Thus, in addition to the moment M and shear V, there is the net axial tension F_3 that must be resisted by the actions developed at the face of the cut. From the equilibrium of the forces on the cut segment:

$$C = F_2 - F_3 \tag{7.5}$$

Hence, the resisting moment at the face of the cut will be:

$$M \sim F_2\, z - F_3\, e \qquad (7.6)$$

where e is the distance between the force F_3 and the centroid of the compression force C. The approximation sign ($\sim$) is used, since the force F_3 acts at the interface between the support and the member, but for the current discussion, it is shown at the centroid of the member, with the restraint modeled as a spring.

In summary, when a member is restrained at supports, the post-tensioning force available to resist the demand moment M is reduced. The amount of reduction, in this example F_3, depends on the relative stiffness of the restraining supports and the post-tensioned member.

The preceding section is a simplification of the mechanism for development of resistance in a post-tensioned member, intended to present the concept. With increase in applied load, there will be an increase in tendon strain, which in turn results in an increase in tendon force. At ultimate limit state (ULS), the force in the tendon (F) is thus $F_2 + \delta F_2$, where δF_2 is the increase in tendon force due to local strain.

In summary, when a member is restrained at supports, the post-tensioning force available to resist the demand moment (M) is reduced. The magnitude of this reduction depends on the relative stiffness of the restraining supports and the post-tensioned member.

7.3.3 Full support restraint

By way of a simple example, the following illustrates the concept and consequence of full support restraint in post-tensioned members. Actual conditions differ, but the concept and phenomenon apply. Figure 7.6 (a) shows a member reinforced with an unbonded tendon and full restraint at its end supports. For presentation of the concept the tendon is assumed straight through the member.

For the condition of full support restraint assumed at stressing, the entire post-tensioning is diverted to the supports, leading to cracks through the depth of the member. Non-prestressed reinforcement helps to control crack width and crack dispersion. To highlight the interaction of prestressing and the restraint of the supports, in this example the contribution of non-prestressed reinforcement is not accounted for.

A. Unbonded tendons: For presentation of concept, self-weight and external loads are not shown. Note that in part (a) of the figure, the tendon retains its force (F_2) across the cracked section, but there is no compressive force on the face of the through crack, since the member is assumed fully fixed against shortening at its end supports. The

162   Post-tensioning in building construction

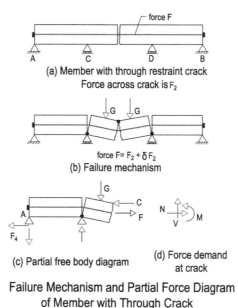

Figure 7.6 Failure mechanism and partial force diagram of member with through crack.

entire tendon force is diverted to the supports A and B. The force F_2 in tendon at service is the same as the restraint of support F_3 (not shown in the figure).

Added load shown in part (b) leads to formation of additional cracks, hinge lines and a failure mechanism.

An idealized partial free body diagram of the left segment of the member for the post-tensioning forces is shown in part (c) of the figure.

The downward displacement of the slab prior to collapse will elongate the tendons along their length, resulting in an increase (δF_2) in the tendon force. The initial tendon force at location of through crack under the service condition (F_2) will increase to its final value F as shown in part (c) of the figure. The impending failure mechanism re-establishes contact between the two sides of the crack, where the compressive force C develops.

For unbonded tendons, the increase in tendon force across the crack prior to failure will be partially transferred to the supports A and B. Although the member itself is restrained against movement, the tendon can slide within its sheathing.

At the ultimate state, the restraint force from the supports increases to F_4. If the member length is longer than is common in building construction, the increase in tendon force can be absorbed by the increase in friction, but this seldom occurs in practical conditions.

Member shortening; precompression: member strength 163

In part (a) of Figure 7.6 the force at restrained support A is F_3. The tendon force at the crack gap is shown as F_2. Since the gap extends through the depth of the member, $F_2 = F_3$. Tendon force through the crack is midified (F_2 is equal to the restraint of the support F_3).

In part (b) the stretching of the tendon increases its force at the crack from the service condition F_2 to F, as shown in part (c).

The force demand (design values) at the crack will be the axial tension N, moment M and shear force V shown in part (d) of the figure.

From equilibrium of forces in the horizontal direction (part c), the axial tension N equals F_4, the restraint of the support at point A at ultimate limit state.

$$N = F_4 \tag{7.7}$$

The demand actions N and M at the cracked section are resisted by the increase in tendon force across the crack. This results in the displacement of the member and compressive force (C) at the newly established contact surface.

The relationships are:

$$C = F - F_4 \tag{7.8}$$

$$M = C\,z \tag{7.9}$$

$$M = (F - F_4)z \tag{7.10}$$

where z is the lever arm between the centroids of the tension and compression forces, and F is the force in the tendon at the crack. Note that the tensile force in the tendon that contributes to the resistance capacity of the cracked section is the difference between the force in the tendon at the crack (F) and the restraint of the support (F_4).

The partial free body diagram of the horizontal forces for the left segment of the member of Figure 7.6 is shown in Figure 7.7. Figure 7.7 illustrates the development of friction force P at ULS. Part (a) shows the segment prior to collapse, where force C develops at contact area between the two sides of the crack. The figure shows the contribution of the friction forces (P) between a strand and its sheathing in developing the compression force (C).

The following shows that the compressive force C that can develop across the crack prior to the collapse of the member is limited to the friction force (P) that builds between a strand and its sheathing (part b of the figure). This is based on the initial premise that the support restraint at A is large enough to prevent the shortening of the member.

From part (b) of the figure:

$$P = F - F_4 \tag{7.11}$$

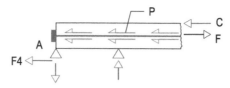

(a) Development of friction force P at ultimate strength

(b) Free body diagram of horizontal forces

Development of Friction Force P at Ultimate Strength

Figure 7.7 Development of friction force P at ultimate strength.

From part (a) of the figure:

$$C = F - F_4 \tag{7.12}$$

where F_4 is the restraint from the support at the ULS. Therefore,

$$C = P \tag{7.13}$$

To arrive at the upper bound for the moment that can develop at the crack, the tendon is assumed to be stretched to its rupture force, recognizing that this is impractical for unbonded tendons, before a member can be considered 'failed.'

The force F in the tendon is calculated as:

$$F = A_{ps} f_{pk} \tag{7.14}$$

where A_{ps} is the tendon cross-sectional area and f_{pk} is its specified strand strength (commonly 1,860 MPa). In practice f_{pk} is unlikely to develop for unbonded tendons. The current discussion is a hypothetical case for an upper-bound limit.

The tendon force F will decrease along the tendon length due to friction between the tendon and its sheathing. For a given tendon profile and friction coefficients, the stress loss due to friction can be calculated with the expressions given in Chapter 8.

Once the friction force P and hence the compressive force C across the crack are determined, the design capacity of the section is known. Note that in an actual structure, the contribution of the non-prestressed steel across the crack contributes to the strength. The compressive force C will be resisted by both the tendons and the non-prestressed reinforcement. The capacity of the section will depend on the location and the magnitude of the tendon force and the location and amount of the non-prestressed reinforcement.

Figure 7.8 shows a member cracked through and broken down into two segments at failure. In the general case, a restraint crack is likely to break the member into two non-equal lengths. For static equilibrium of the member shown, the restraining forces (F_4) on each side of the crack must be equal. Thus, the friction force (P) that can be sustained across the crack is that from the segment with the smaller friction loss – typically the shorter side of the member. Concluding with $C = P$, the moment capacity is:

$$M = Pz \qquad (7.15)$$

In summary, for the conditions discussed, the maximum tensile force that will be available to develop the resisting moment at the crack is limited to the friction that develops between the tendon and its sheathing at ULS. Figure 7.9 illustrates the development of tendon resisting force at ULS of member reinforced with unbonded tendons.

In Figure 7.9, for the service condition $F_2 = F_3$. This is the condition prior to the application of added load and formation of the compression force C (refer to Figure 7.6). F_u is the ultimate force developed in the 'unbonded' tendon. In part (b) the tendon force in service F_2 at the location of crack is equal to the support restraint F_3. At strength limit, the tendon force at location of crack is increased to F_u.

In the preceding diagram, the force ($F_u - F_4$) is the force that will be available to resist applied moments at ULS – the moment capacity of the section. The force ($F_u - F_4$) is the friction force between the strand and its sheathing at ULS.

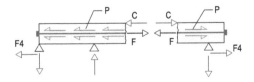

Partial Free Body Diagram of a
Non-Symmetrical Member Cracking

Figure. 7.8 Partial free body diagram; nonsymmetrical cracking.

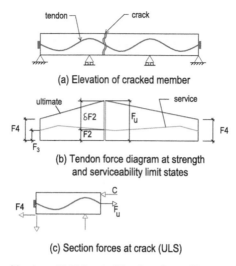

Figure 7.9 Member with unbonded tendons.

Further, it is concluded that when members reinforced with unbonded tendons experience excessive support restraints, the friction between tendon and its sheathing plays a significant role in the strength available from the tendon at the member's ultimate strength. The larger the friction force between a tendon and its sheathing, the greater will be the available tendon strength to resist applied loads.

B. Bonded tendons: Members reinforced with bonded tendons develop larger moment capacity at locations of restraint cracks compared to members that are reinforced with the same amount of unbonded post-tensioning. There are three reasons.

First, bonded tendons typically develop their specified strength (f_{pk}) prior to failure, whereas members reinforced with unbonded tendons tend to undergo large deflections, and reach failure due to crushing of concrete or excessive deflection, before tendons reach their specified strength. Consequently, EC2 and ACI 318 specify a significantly lower permissible stress for unbonded than bonded tendons for flexural capacity computations of post-tensioned members.

Depending on the span dimensions for unbonded tendons, ACI 318 limits the increase in tendon stress at ultimate strength to between 206 and 413 MPa,[1] whereas in EC2 the increase is limited to merely 100 MPa.[2] This is about a 7% to 9% gain in strength over the service condition, leaving about 30% of a tendon's strength untapped at member failure.

Second, for members reinforced with bonded tendons, the increase in demand moment at a point results in an increase in the tendon force at the same location. This local increase in tendon force is not compromised by the restraint of the supports. On the other hand, for unbonded tendons – as outlined in the previous section – support restraints can diminish the effectiveness of the local increase in tendon force in resisting an applied moment. This is detailed in the following.

Third, compared to unbonded tendons, for bonded tendons the higher friction force between the strand and the sheathing at stressing works advantageously at the strength limit state of a cracked section.

Consider the member with a bonded tendon. Figure 7.10 shows the distribution force in tendon at ULS.

Let the restraint from the supports be large enough to cause cracking as shown in part (a). The force in the tendon at the time of grouting follows essentially the friction diagram shown in part (b). Let the force in the tendon at location of crack in the service condition be F_2. For the static equilibrium of the arrangement shown, F_2 is equal to the restraint of the support (F_3) while the gap at the crack is open. An increase in the applied moment

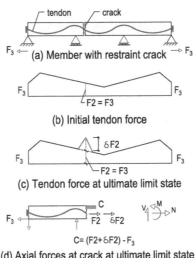

Figure 7.10 Member with bonded tendon and restraint crack; forces at strength limit.

at the crack location will tend to elongate the tendon locally leading to an increase in the tendon force by δF_2 (part c of the figure).

The demand actions at the location of crack (part d of the figure) are moment M, axial force N and shear V. From equilibrium of the forces, N is equal to F_3, the force due to restraint of the support.

The tensile force available to resist the demand actions at the crack location is:

$$T = F_2 + \delta F_2 - F_3 \tag{7.16}$$

Since at location of crack $F_2 = F_3$, the available force (T) to resist the induced moment will be equal to δF_2.

Likewise, from equilibrium of forces the compression force C is:

$$C = (F_2 + \delta F_2) - F_3 = \delta F_2 \tag{7.17}$$

The moment that can be developed at the crack M is equal to:

$$M = Cz = \delta F_2 z \tag{7.18}$$

C. Comparison between unbonded and bonded systems: Figure 7.11 compares the performance of a member reinforced with unbonded post-tensioning to that of a member with bonded post-tensioning with respect to the impact of restraint cracks on member's bending

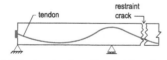

(a) Member elevation with restraint crack

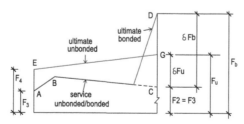

(b) Tendon force at service and ultimate limit states

Available Resisting Force at Ultimate Limit State

Figure. 7.11 Tendon force at ultimate limit state for bonded and unbonded post-tensioning systems.

strength. The diagram shows the distribution of force in the tendon for the segment to the left of the crack.

Line ABC is the force in the tendon at service condition. It is influenced by the friction and seating loss at stressing. To illustrate the concept, it is assumed to be the same for bonded and unbonded systems at service condition. For the ultimate moment at the crack location there will be a local increase in tendon force for the bonded system marked by point D (stress F_b). The tendon force available to resist the demand moment is equal to the local increase in tendon force (δF_b) shown by CD. At strength limit, the distribution of force in the unbonded tendon will be governed by the re-alignment of the friction force along the tendon length from line ABC to line EG. The available force to resist the moment demand will be

$$(F_2 + \delta F_u) - F_4 = F_u - F_4 \tag{7.19}$$

Referring to Figure 7.11, the net force (T) from the post-tensioning tendons available at the crack to resist the demand moment is:

Unbonded: $T = F_u - F_4$ (7.20)

Bonded: $T = F_b - F_3 = \delta F_b$ (7.21)

7.4 EXAMPLE

Figure 7.12 shows the reflected ceiling of a first elevated post-tensioned slab supported on perimeter walls with through cracks extending over the entire width of the floor. Post-tensioned tendons are grouped over the lines of

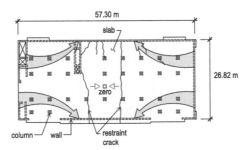

View of Reflected Ceiling Showing
Cracks in PT Slab

Figure 7.12 Reflected ceiling plan of a post-tensioned slab on perimeter walls with extensive restraint cracks.

columns in the long direction. The tendons are distributed uniformly in the short direction. The slab has developed multiple long restraint cracks extending through its depth. The restraining walls on the long sides interrupt the precompression in the slab leading to no compression from the tendons through the central region of the slab.

The two long cracks in the transverse direction that extend through the entire width of the slab indicate that the precompression from post-tensioning in the long direction of the slab has been fully diverted to the long perimeter walls.

There is no precompression from prestressing in the central slab region between the long through cracks. Tendons pass through this region, however. Tendons provide uplift (lateral force) determined by their profile and force, but do not contribute to the moment capacity of the central region between the through cracks. Evidently, the safety factor of the slab over the central region is reduced.

7.5 PRECOMPRESSION IN MULTI-STORY BUILDINGS

7.5.1 Inter-story redistribution of precompression

Precompression in post-tensioned multi-story buildings and its impact on the floor's strength differ from the scenario explained above for the first elevated slab.

In multi-story frames, the floors are commonly constructed and stressed sequentially with progress in construction. Figure 7.13 shows the idealized

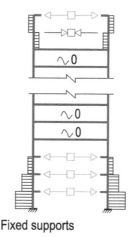

Fixed supports

Figure 7.13 Idealized elevation of a multi-story post-tensioned building showing loss or gain in precompression from prestressing of each level.

elevation of a multi-story post-tensioned building illustrating the loss or gain in precompression from prestressing of each level. At lower levels a fraction of precompression is lost to the supports. The loss results in shear forces shown in the walls. At the roof level there is loss of precompression accompanied by gain in precompression to the level below the roof.

At application of prestressing to the first elevated floor, the second floor is generally not present. The restraint of the supports and foundation absorb a fraction of the precompression intended for the floor being stressed.

When the second floor is post-tensioned, the restraint of its supports is somewhat less than that experienced by the floor below. Again, a fraction of the precompression of the new floor is diverted to the structure below it. This results in partial recovery of loss of prestressing in the first floor.

With progress in construction, at a typical upper level, the short-term difference in the shortening of a floor being stressed and the floor immediately below is the shortening that takes place in the level below during the construction cycle of a floor.

For typical buildings, the construction cycle can be about 7 days. There will be an initial loss of precompression to the levels below, but the loss will be recovered when the level above is constructed and stressed.

The pattern continues, until the uppermost level is stressed. The precompression lost to the penultimate floor from the uppermost roof level is not recovered, since there is no floor above it.

With lapse of time, over a period of 2 years or more, shortening due to creep, shrinkage, relaxation in prestressing and the effect of aging in concrete between a typical upper level and one above or below erodes the long-term difference of precompression between adjacent identical levels of multi-story buildings.

The preceding description is valid except for the lowest levels, where the loss of precompression to the foundation will change over time, but will not be eliminated, as is the case for the upper levels.

7.5.2 Impact of stiff walls at interior of floor slab

Interior single or compound walls, such as elevator core walls, do not result in long-term loss of precompression from the slab to the walls over the unsupported front and back regions of a wall. In other words, the floor regions at the front and back of the walls receive full precompression from prestressing.

Figure 7.14 shows the partial elevation and plan of a multi-story frame with interior walls. Part (a) of the figure indicates that for a typical level, the precompression (P) from prestressing spreads into the wall, but it is recovered by the slab beyond the wall.

The strength design of the slab for the regions on the sides of the walls is not as critical as the regions at the front and the back of the wall ends, since the regions on the sides of the walls benefit directly from the wall support.

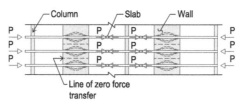

(a) Partial elevation - flow of precompression

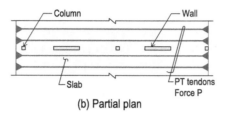

(b) Partial plan

Precompression from Post-Tensioning at Upper Levels of Multistory Building

Figure 7.14 Partial plan of an upper level of a multi-story frame showing the precompression in the floor.

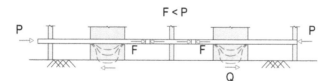

Diversion of Post-tensioning From First Elevated Floor

Figure 7.15 Elevation of a post-tensioned first-level floor; part of the precompression from prestressing is diverted to the foundation.

The preceding conclusion regarding the restraining effects of the walls does not apply to the first elevated floor over the foundation. The precompression lost to the foundation level is not recovered in full. Figure 7.15 shows the elevation of a post-tensioned first-level floor. It illustrates the partial loss of precompression to the foundation.

7.6 LONG-TERM SHORTENING

In addition to immediate shortening under jacking force, a post-tensioned member continues to shorten with time, albeit at a reducing rate over time.

It is important for the shortening to take place unhindered – as much as practical – in order to mitigate the member precompression to be diverted to adjoining restraining elements.

The shortening of a post-tensioned member is primarily the result of change in concrete volume from shrinkage, creep, elastic shortening and temperature change.

Chapter 8 includes a section on the estimate of shortening in post-tensioned members. A detailed account of shortening calculations and a numerical example is given in reference [Aalami, B. O., 2021].

NOTES

1. ACI 318-10, 20.3.2.4.1.
2. EC2, EN 1992-1-1:2004, 5.10.8 (2).

REFERENCES

Aalami, B. O. (2021), *Post-Tensioning; Concepts; Design; Construction*, PT Structures Inc., Palo Alto, www.PTStructures.com, pp. 570.

Aalami, B. O. (2018), "Support Restraints and Strength of Post-Tensioned Members, Part 1," *Structure Magazine*, www.StructureMag.org, January 2018, pp. 18–20.

Aalami, B. O. (2017), "Support Restraints and Strength of Post-Tensioned Members, Part 2," *Structure Magazine*, www.StructureMag.org, October 2017, pp. 35–36.

Leonhardt, F. (1964), *Prestressed Concrete Design and Construction*, Wilhelm Ernst & Sohn, Berlin, München, 667 pp.

Chapter 8

Stress losses in post-tensioning

8.1 STRESS LOSSES

Post-tensioning tendons are typically pulled to 80% to 85% of their ultimate strength before being locked at the face of the member. The initial stress at jacking force will be partially lost, partly at stressing and partly with passage of time. The initial stress loss at stressing is referred to as immediate loss. The loss over time is the long-term loss.

The stress loss results in the effective tendon force available for service of the post-tensioned member to be less than the force imparted at time of jacking. The long-term 'effective force' is typically used in design of the service design of the member.

In common construction of residential and commercial buildings using the commonly available prestressing strands and hardware, the rough estimate of the stress condition for preliminary designs is:

Strength of prestressing strand   1,860 MPa
Jacking stress   1,488 MPa

Effective stress after all losses

- Bonded tendons   1,100 MPa
- Unbonded tendons 1,200 MPa

This chapter explains the causes of stress loss and a method for its calculation. The stress losses fall into two categories, namely immediate losses and long-term losses. The immediate losses take place during the jacking process and locking of the strand. The long-term losses follow over time after the stand is seated.

The pull force at the jacking end drops along the length of the tendon from friction between the strand and its sheathing. The 'friction' is the primary factor of immediate stress loss.

Once pulled to the design force, the post-tensioning tendon is locked in position at the pulling end. The locking device commonly consists of two conical wedges that grip the strand. At release of pulling force, the wedges retract into the anchor cone cavity to grip-lock the strand. The draw-in of the wedges into the anchor device to secure the pulled strand results in local loss of a fraction of stress in the pulled strand. Figure 8.1 shows the loss in tendon force along its length and the loss from retraction of the wedges in locking the strand.

For a long strand the loss of force at the far end can be significant. To partially recover the loss of force, in some instances the strand is pulled at the other end subsequent to the completion of stressing at the first end. Figure 8.2 shows the distribution of force along the tendon length from stressing of the tendon at the second end. This results in a more favorable force distribution in the strand.

In construction practice many engineers require the tendons longer than 35 m to 40 m to be stressed at both ends, in order to maintain the effective tendon stress for design at the level stated above.

Figure 8.2(e) shows conceptually the loss of stress with time and the residual 'effective' stress for design of the member in service for a strand that is pulled at both ends.

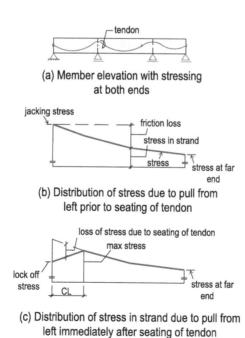

Figure 8.1 Tendon stressing at one end and the distribution of force in tendon.

Stress losses in post-tensioning   177

(d) Stress in strand due to pull from right prior to seating of tendon

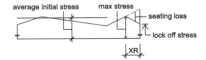

(e) Stress in strand immediately after seating of tendon at the right end

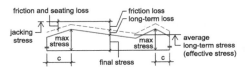

(f) Distribution of final stress in strand after immediate and long-term losses

Friction and Long Term Stress Loss Diagram

Figure 8.2 Tendon stressing at second end and the distribution of force in tendon.

8.2 IMMEDIATE LOSSES

The immediate losses are from (i) friction between the strand and its sheathing and (ii) the retraction of the strand in being seated.

8.2.1 Friction

The stress at any point along a strand is related to the jacking stress by the following expression.

$$P_x = P_j e^{-(\mu\alpha + Kx)} \tag{8.1}$$

where

P_x = stress at distance x from the jacking point;
P_j = stress at jacking point;
μ = coefficient of angular friction;
A = total angle change of the strand in radians from the stressing point to distance x;

X = distance from the stressing point; and
K = wobble coefficient of friction expressed in radians per unit length of strand (rad/unit length).

Figure 8.3 illustrates for a tendon segment the design-intended change in angle α and the unintended change of angle β due to workmanship. The latter is a function of construction practice. The unintended change in angle is given by the wobble coefficient (K).

The friction coefficients for the common seven-wire unbonded and bonded tendons are:

Unbonded tendons:
$\mu = 0.07$
$K = 0.0047$ rad/m

Bonded tendons – sheet metal ducts housing up to five strands:
$\mu = 0.25$
$K = 0.001$ rad/m

8.2.2 Elongation

Jacking force at the first end results in elongation given by the expression 8.2.

$$\Delta = \int \frac{P_x}{AE_s} dx \tag{8.2}$$

where

A = cross-sectional area of the tendon;
dx = element of length along tendon;

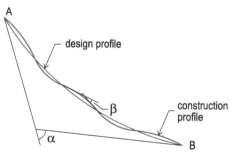

α = intended angle change
β = unintended angle change

Angle Change in Tendon Profile

Figure 8.3 Angle change in tendon profile by design and workmanship.

E_s = modulus of elasticity of the prestressing steel (typically taken as 195,000 MPa);
P_x = tendon force at distance x from the jacking end; and
Δ = calculated elongation.

The elongation is counteracted by friction between the strand and its sheathing or duct.

8.2.3 Seating (draw-in) losses

After seating the tendon, the force in the tendon will drop locally. Figure 8.4 shows the idealized tendon stress diagram with anchor set influence. The extent of the anchor set influence is shown by distance 'c' in the figure.

The shape of the tendon's final stress over distance 'c' is the mirror image of the force in the tendon prior to seating.

The area of the triangle formed between the stress curve before and after seating of the tendon is equal to the distance of the strand's draw-in at seating – typically 6 mm for common strands.

The distance 'c' can be calculated from the known area of the stress loss triangle shown in the figure. In the general case, narrow vertical strips and integration is used. But, if the friction loss is idealized as a straight line as shown in the figure, the anchor loss distance 'c' is given by expression (8.3).

$$c = \sqrt{\frac{E\delta L}{12d}} \qquad (8.3)$$

where

c = anchor set distance in mm;
d = friction loss over distance L in MPa;

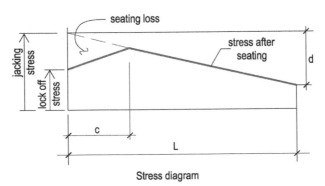

Anchor Set Influence Distance Diagram

Figure 8.4 Idealized tendon stress diagram showing anchor set influence distance c.

E = modulus of elasticity of the strand MPa; and
L = tendon length in mm.

8.3 LONG-TERM STRESS LOSSES

A detailed account of long-term stress losses is given in references [PCI, 2017; Zia., P., et al., 1979, Aalami, B. O., 2014].

The long-term loss is given by:

$$TL = ES + CR + SH + RE \tag{8.4}$$

where

ES = stress change due to elastic deformation;
CR = stress loss due to creep;
RE = stress loss due to relaxation in prestressing steel;
SH = stress loss due to shrinkage of concrete; and
TL = total loss of stress.

For common construction, as an alternative to lump sum stress loss assumption, simplified expressions outlined in the following are often used [ACI-ASCE1979].

8.3.1 Elastic shortening of concrete (ES)

Elastic shortening effect is the loss in tendon force that has already been stressed and seated from shortening of the member from subsequent application of the post-tensioning force in other tendons of the structure.

If there is only one tendon in a member, there will be no loss from elastic shortening since the elastic shortening will have occurred before the tendon is locked in place.

Generally, there is more than one tendon in a member, specifically in floor slabs. As each tendon is tensioned, there will be loss of prestress in the previously tensioned tendons due to the elastic shortening of the member under the newly stressed tendon.

The expression for elastic shortening at stressing is given by:

$$ES = \left(\frac{E_s}{E_{ci}}\right) f_{cir} \tag{8.5}$$

where

E_s = the elastic modulus of the prestressing steel;
E_{ci} = the elastic modulus of the concrete at time of prestress transfer; and
f_{cir} = net compressive stress in concrete at center of gravity of tendons after prestress has been applied to concrete.

Stress losses in post-tensioning 181

For bonded tendons, at stressing, the ducts in which bonded tendons are housed are not grouted. Thus, the elastic shortening equations for unbonded tendons applies to bonded tendons too.

For bonded tendons, subsequent to grouting, there will be additional local elastic change in the tendon due to flexing of the member and strain compatibility between the tendon and its surroundings. The local strain due to flexing of the members impacts the stress in the tendon. A numerical example at the end of this chapter illustrates the condition.

8.3.2 Creep of concrete (CR)

Over time, compression from post-tensioning shortens the member. Creep is shortening of the member under sustained compression.

For members reinforced with unbonded tendons, the stress loss due to creep is:

$$CR = K_{cr}\left(\frac{E_s}{E_c}\right)f_{cpa} \qquad (8.6)$$

where

E_c = elastic modulus of the concrete at 28 days;
E_s = elastic modulus of the prestressing steel;
f_{cpa} = average member precompression from prestressing; and
K_{cr} = maximum creep coefficient; 2.0 for normal weight concrete.

For bonded tendons, the change in tendon stress due to creep is a function of the local stress adjacent to the tendon. The numerical example at the end of this chapter illustrates the application.

8.3.3 Shrinkage of concrete (SH)

Shrinkage is mostly due to loss of water from concrete. Loss of water is influenced by the member's volume/surface ratio, and the ambient relative humidity.

The effective shrinkage strain, ε_{sh} is adjusted for the member's volume-to-surface ratio and the ambient relative humidity.

The common shrinkage strain for normal conditions is taken as 550×10^{-6}. The shrinkage strain for specific conditions can be estimated from expression (8.7)

$$\varepsilon_{sh} = 8.2 \times 10^{-6}\left(1 - 0.00236\frac{V}{S}\right)(100 - RH) \qquad (8.7)$$

The stress loss due to shrinkage is:

$$SH = 8.2 \times 10^{-6} K_{sh} E_s \left(1 - 0.00236 \frac{V}{S}\right)(100 - RH) \tag{8.8}$$

where

K_{sh} = factor that accounts for the shrinkage that would have taken place before the prestressing is applied;
RH = relative ambient humidity (%); and
V/S = volume-to-surface ratio of the member.
K_{sh} is given in Table 8.1.

In structures that are not moist-cured, K_{sh} is typically based on the time when the concrete was cast. In most structures, the prestressing is applied within 5 days of casting the concrete, whether or not it is moist-cured.

If the ultimate shrinkage value of the concrete is different from the 550 micro-strains used above, the calculated stress loss must be adjusted by prorating the shrinkage coefficient SH.

8.3.4 Relaxation of prestressing steel (RE)

Relaxation is the gradual decrease of stress in material under constant strain. In the case of steel, it is the result of a permanent alteration of the grain structure. The rate of relaxation at any point in time depends on the stress level in the prestressing steel at that time. Because of other prestress losses, there is a continual reduction of the tendon stress which causes a corresponding reduction in the relaxation rate.

The prestress loss due to relaxation of the tendons is:

$$RE = \left[K_{re} - J(SH + CR + ES)\right]C \tag{8.9}$$

where K_{re} and J are functions of the type of steel. C is a function of both the type of steel and the initial stress level in the tendon.

The coefficients for the common 1,860 MPa stress relieved strands used in industry are:

K_{re} = 34.50 MPa
J = 0.04
C = 1.28 for f_{pi}/f_{pk} = 0.8

Table 8.1 Shrinkage coefficient K_{sh}

Days *	1	3	5	7	10	20	30	60
K_{sh}	0.92	0.85	0.80	0.77	0.73	0.64	0.58	0.45

*Days refer to the time from the end of moist-curing to the application of prestressing.

For stressing more than 60 days after curing, assume 0.45.

Coefficients for other strand types and conditions are listed in reference [Aalami, B. A., 2014].

8.4 EXAMPLES

8.4.1 Friction loss calculation

GIVEN

Figure 8.5 shows a four-span slab over several parallel beams. The parallel beams are shown as point supports in the figure. The slab is post-tensioned with unbonded tendons providing an average precompression of 1.70 MPa after all losses.

The tendon profile for the first half of the first span and the second half of the last span is a simple parabola. The remainder of the tendon length is a reversed parabola between each low and high tendon point.

The tendon is stressed at the left end of the slab only.

REQUIRED

Calculate the friction loss of the tendon and its long-term stress losses.
 Geometry

   Four continuous spans each 5.50 m
   Slab thickness 120 mm

Material properties
 Concrete:

Compressive strength (28 days)	= 28 MPa
Weight	= 2,400 kg/m^3
Modulus of elasticity	= 24,850 MPa
Age of concrete at stressing	= 3 days
Compressive strength at stressing, f'_{ci}	= 13 MPa

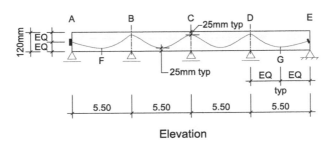

Four Span Slab Stressed at One End

Figure 8.5 Elevation of prestressed member.

Prestressing:

Low relaxation, unbonded system	
Strand diameter	= 13 mm
Strand area	= 99 mm²
Modulus of elasticity	= 200,000 MPa
Coefficient of angular friction, μ	= 0.07
Coefficient of wobble friction, K	= 0.0046 rad/m
Ultimate strength of strand, f_{pk}	= 1,860 MPa
Ratio of jacking stress to strand's ultimate strength	= $0.8 f_{pk}$
Anchor set	= 6 mm
Average precompression	= 1.7 MPa

The friction loss between two points along the tendon length among other factors depends on the angle change between the two points. Figure 8.6 shows the angle change for tendon segments that follow parabolic curves.

Using the preceding figure as guide and the friction loss formula, the loss in stress along the tendon can be readily calculated. The outcome is summarized in Table 8.2. The calculated stress at the far end is 0.863 times the jacking stress.

The drop in stress at the far end 'd' using Table 8.2 is given by:

$d = (1 - 0.863) \times 0.8 \times 1,860 = 203.856$ MPa

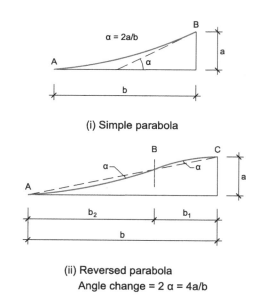

Figure 8.6 Angle change for parabolic segments.

Table 8.2 Friction loss values for the example member

Segment	Length m	a mm	b mm	Angle radians (α)	Total angle radians (α)	x m	Point	μα + Kx	$e^{-(\mu\alpha + Kx)}$
AF	2.75	35	2750	0.0254	0.0254	2.75	F	0.014428	0.986
FB	2.75	70	2750	0.10181	0.1273	5.50	B	0.03420	0.966
BC	5.50	70	5500	0.2036	0.3054	11.00	C	0.06595	0.936
CD	5.50	70	5500	0.2036	0.5090	16.50	D	0.1134	0.893
DG	2.75	70	2750	0.10181	0.6018	19.25	G	0.1331	0.875
GE	2.75	35	2750	0.0254	0.6363	22.00	E	0.1475	0.863

The anchor set influence distance 'c' for the specified 6 mm seating loss from the expression 8.3 is given by:

$$c = [200{,}000 \times 6 \times 22{,}000/(12 \times 203.856)]^{0.5} = 3{,}285 \text{ mm}$$

The elongation at stressing is given by the area below the stressing diagram. In common engineering calculation the drop in stress along the tendon is assumed linear.

Elongation = $0.5 \times 1{,}860 \times [0.8 + 0.8 \times 0.863] \times 22{,}000/200{,}000 = 152$ mm

Elongation after anchor set = 152 - 6 = 146 mm

8.4.2 Long-term loss calculation of member with unbonded tendons

GIVEN

The four-span slab of the preceding example is part of a beam and slab construction. The slab spans between the equally spaced parallel beams.

REQUIRED

Calculate the long-term stress losses in the slab's prestressing tendons.

Long-term loss

Long-term loss calculation for first span:

(i) Elastic shortening from expression 8.6:

$$ES = 0.5 \left(\frac{E_s}{E_{ci}}\right) f_{cpa}$$

f_{cpa} = 1.70 MPa
E_{ci} = 16,800[1] MPa
E_s = 193,000 MPa
ES = 0.5 (193,000/16,800)1.70 = 9.76 MPa

(ii) Shrinkage of concrete from expression 8.8:

$$SH = 8.2 \times 10^{-6} K_{sh} E_s \left(1 - 0.00236 \frac{V}{S}\right)(100 - RH)$$

V/S = 60 mm
RH = 80 %
K_{sh} = 0.85 (from Table 8.1 for 3 days)
SH = 8.2×10⁻⁶×0.85×193,000(1- 0.00236×60) × (100-80) = 23.09 MPa

(iii) Creep of concrete from expression 8.6:

$$CR = K_{cr} \left(\frac{E_s}{E_c}\right) f_{cpa}$$

E_c = 24,850 MPa
CR = 2× (193,000/24,810) ×1.70 = 26.45 MPa

(iv) Relaxation of strands from expression 8.9:

$$RE = \left[K_{re} - J(SH + CR + ES)\right]C$$

For 1,860 MPa low relaxation strand
K_{re} = 34.50 MPa
J = 0.04
C = 1.28 for f_{pi}/f_{pu} = 0.8
RE = [34.50 - 0.04(23.09 + 26.45 + 9.76)]1.28 = 41.12 MPa

TL = ES + CR + SH + RE

Total stress losses = 9.76 + 26.45 + 23.09 + 41.12 = 100.42 MPa

The long-term losses are summarized in Table 8.3.

Table 8.3 Summary of long-term stress losses

Stress Loss Item	MPa	%
Elastic shortening (ES)	9.76	10
Creep (CR)	26.45	26
Shrinkage (SH)	23.09	23
Relaxation (RE)	41.12	41
Total	100.42	100

8.4.3 Long-term loss calculation of member reinforced with bonded tendons

The friction and elongation calculation procedures are identical to that of unbonded systems.

The long-term loss calculation differs significantly, however. The stress loss at a point in the bonded tendon is tied to the strain in concrete adjacent to the tendon at the same location. Consequently, for detailed design, stress loss must be calculated at each critical location along the tendon length.

GIVEN

Geometry

Figure 8.7 shows the elevation of a two-span flanged beam post-tensioned with bonded tendons. The section of the beam is shown in Figure 8.8.
Material properties
Concrete:

Compressive 28-day strength	= 27.58 MPa
Weight	= 2,400 kg/m³
Modulus of elasticity	= 24,849 MPa
Age of concrete at stressing	= 3 days
Compressive strength at stressing, $f_{ck,i}$	=20 MPa

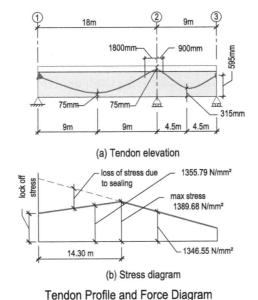

Figure 8.7 Tendon profile and force diagram.

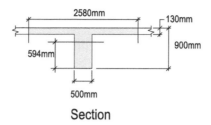

Figure 8.8 Section of the member.

Prestressing:

Post-tensioning is with a multi-strand tendon consisting of 8–13 mm diameter low-relaxation strands stressed at one end.

Strand diameter	= 13 mm
Strand area	= 98 mm²
Modulus of elasticity	= 193,000 MPa
Coefficient of angular friction μ	= 0.20
Coefficient of wobble friction K	= 0.0002 rad/m
Ultimate strength of strand f_{pk}	= 1,860 MPa
Ratio of jacking stress to strand's ultimate strength	= 0.8
Anchor set	= 6 mm

Loads

Self-weight of beam for 6 m tributary = 26 kN/m
Superimposed sustained load in addition to self-weight after the beam is placed in service = 1.02 kN/m² = 2.63 kN/m

Live load is generally not considered for the long-term stress loss in bonded tendons.

REQUIRED

Calculate the long-term stress losses in the prestressing tendons at midpoint of the first span and over the interior support.

The long-term stress loss calculation consists of the following steps:

1. Determine the initial stress in the tendon at midspan and over the second support.
2. Determine the bending moments and stresses at midspan and over the first support from self-weight and the superimposed sustained loads. Also, calculate stresses due to post-tensioning tendons.
3. Use the applicable relationships to calculate the long-term stress losses at the locations selected.

A. Calculation of initial stress in tendon

The initial stress in the tendon is determined from the friction loss relationship outlined earlier in this chapter. The calculated tendon stress at jacking is:

At midspan f_{pi} = 1,355.79 MPa
At support f_{pi} = 1,346.55 MPa

B. Moments and stresses

The section properties of the beam are:
Cross sectional area $A = 720{,}400$ mm^2
Second moment of area $I = 5.579 \times 10^{10}$ mm^4

Neutral axis to bottom fiber $Y_b = 594.50$ mm
Neutral axis to top fiber $Y_t = 305.50$ mm

Distance of neutral axis to tendon centroid

At midspan $c = 594.50 - 75 = 519.50$ mm
At support $c = 305.50 - 75 = 230.50$ mm

The distribution of self-weight moments of the beam is obtained using standard plane frame software. Figure 8.9 shows the balanced loads from post-tensioning and post-tensioning moments.

The combined moments from the quasi-permanent load combination are:

Moment at midspan   658.34 kNm
Moment at support   -789.29 kNm

The stresses in concrete at the center of gravity of the tendon (f_g) due to weight of the structure at time of stressing is calculated next. This is a hypothetical point for concrete, as there is no concrete at CGS of tendons. The value will be used to determine the strain in strand at the location.

Stresses from post-tensioning and self-weight are contributary to long-term changes in tendon stress. The calculated stresses in concrete at the location of tendons are:

Stress from self-weight (f_g) at midspan: 6.14 MPa tension
Stress from self-weight (f_g) at support: 3.25 MPa tension

Stresses due to superimposed sustained load (f_{cds}) may be prorated from the self-weight stresses:

190   Post-tensioning in building construction

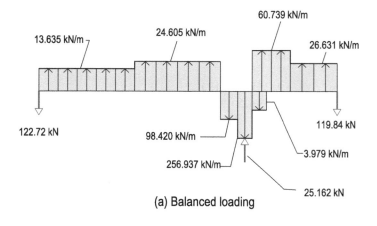

(a) Balanced loading

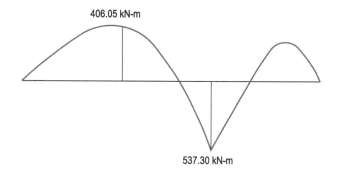

(b) Post-tensioning moments

Balanced Loads and Moments

Figure 8.9 Post-tensioning balanced loads and moments.

The initial stress in concrete is calculated from the balanced loads immediately after the tendon is seated – before long-term losses take place. Figure 8.9 gives the values of the balanced loads and the associated post-tensioning moments.

The balanced moments from post-tensioning area:

At midspan $M_b = -406.05$ kNm
At support $M_b = 537.30$ kNm

Initial concrete stress due to post-tensioning f_{cpi}:
  At midspan:

The initial post-tensioning force is:
$P_{pi} = 8 \times 98 \times 1{,}355.79/1{,}000 = 1{,}062.94$ kN

$$f_{cpi} = \left(\frac{P_{pi}}{A} + \frac{M_b}{I}c\right) = \left(\frac{1{,}062.94 \times 10^3}{72{,}0400} + \frac{406.05 \times 10^6 \times 519.50}{5.579 \times 10^{10}}\right)$$

$$= 5.26 \text{ MPa C } (compression)$$

At support:
The initial post-tensioning force is:
$P_{pi} = 8 \times 98 \times 1{,}346.55/1{,}000 = 1{,}055.70$ kN

$$f_{cpi} = \left(\frac{1{,}055.70 \times 10^3}{720{,}400} + \frac{537.30 \times 10^6 \times 230.50}{5.579 \times 10^{10}}\right)$$

$$= 3.68 \text{ MPa C } (compression)$$

C. Long-term stress loss
At midspan
(i) Elastic shortening from expression 8.5:

$$ES = K_{es}\left(\frac{E_s}{E_{ci}}\right)f_{cir}$$

$K_{es} = 0$ (all strands are pulled and anchored simultaneously); hence, $ES = 0$ MPa

Observe that, in this example, the long-term losses due to elastic shortening are zero since all the strands are stressed and anchored simultaneously.

(ii) Creep of concrete:

For the calculation of losses due to creep, the initial stress in concrete f_{cir} will be calculated with both the self-weight and the sustained superimposed loadings considered as active. Hence, the net sustained stress from the actions of post-tensioning and self-weight are:
$f_{cir} = 5.26 - 6.14 = -0.88$ MPa (Tension)
Sustained stress from the action of other loads f_{cds} is:
$f_{cds} = 0.62$ MPa (Tension) from stress calculations

$$CR = K_{cr}\left(\frac{E_s}{E_c}\right)(f_{cir} - f_{cds}) \tag{8.10}$$

$K_{cr} = 1.6$
$E_c = 4{,}700 \times 27.58^{1/2} = 24{,}683$ MPa

$(f_{cir} - f_{cds}) = -0.88 - 0.62 = -1.50$ MPa (Tension)

Observe that the net stress $(f_{cir} - f_{cds})$ is tensile. Stress loss due to creep is associated with compressive stresses only. A negative sum is substituted by zero. The tension results in an increase in tendon stress, but the change is assumed as zero.

$CR = 1.6 \times (193{,}000/24{,}683) \times 0 = 0$ MPa

(iii) Shrinkage of concrete from expression 8.8:

The relationship for shrinkage is:

$$SH = 8.2 \times 10^{-6} K_{sh} E_s \left(1 - 0.00236 \frac{V}{S}\right)(100 - RH)$$

where

$K_{sh} = 0.85$ (for stressing at day 3 – Table 8.1)
$RH = 70$ (given relative humidity)
V/S = volume to surface ratio = $(2{,}580 \times 130 + 500 \times 770)/(2 \times 2{,}580 + 2 \times 770) = 107.52$ mm
$SH = 8.2 \times 10^{-6} \times 0.85 \times 193{,}000(1 - 0.00236 \times 107.52)(100-70) = 30.12$ MPa

(iv) Relaxation of strands:

$$RE = \left[K_{re} - J(SH + CR + ES)\right]C$$

$f_{pi} = 1{,}355.79$ MPa
$f_{pi}/f_{pk} = 1{,}355.79/1{,}860 = 0.73$
$C = 1.28$
$K_{re} = 34.50$ MPa
$J = 0.04$
$RE = [34.50 - 0.04 \times (30.12 + 0 + 0)] \times 1.28 = 42.61$ MPa

(v) Total loss:

$TL = 0 + 0 + 30.12 + 42.61 = 72.73$ MPa

At second support

Over the second support the stress losses are computed as follows:

(i) Elastic shortening from expression 8.5:

$$ES = K_{es}\left(\frac{E_s}{E_{ci}}\right)f_{cir}$$

$K_{es} = 0$ (all strands are pulled and anchored simultaneously)
Hence, $ES = 0$ MPa

(ii) Creep of concrete:

The creep is the function of long-term stress. The long-term stress is given by the combination of stress in prestressing steel, stress due to

Stress losses in post-tensioning 193

self-weight and superimposed dead load plus the live load from quasi-permanent load combination.

Stress in concrete from prestressing f_{cpi}:
f_{cpi} = 3.68 compression
Stress due to self-weight and superimposed dead load.
In this case the stress is essentially from self-weight (f_g) at support:
3.25 tension
f_g = 3.25 tension

Stress from quasi-permanent live load (sustained stress) needs to be included in the creep estimate. In this case, its value is not significant.

Net long-term stress = 3.68 − 3.25 = 0.43 compression

$$CR = K_{cr}\left(\frac{E_s}{E_c}\right)(f_{cir} - f_{cds})$$

CR = 1.60(193,000/24,683) (0.43-0.33) =1.25 MPa

(iii) Shrinkage of concrete:
Same as in the midspan
SH = 30.12 MPa

(iv) Relaxation of Strands from Expression 8.9:

$$RE = \left[K_{re} - J(SH + CR + ES)\right]C$$

f_{pi} = 1,346.55 MPa
f_{pi}/f_{pk} = 1,346.55/1,860 = 0.72 < 0.80

For values of $f_{pi}/f_{pu} < 0.80$ the coefficient C will be less than 1.28 quoted herein. It is 0.85 [Aalami, 2014]. Conservatively 1.28 is used. This overestimates the stress relaxation losses.

C = 1.28
K_{re} = 34.50 MPa
J = 0.04
RE = [34.50 − 0.04(30.12 + 1.25 + 0)] ×1.28 = 42.55 MPa

(v) Total Loss from Expression 8.4:

$$TL = ES + CR + SH + RE$$

$$TL = 0 + 1.25 + 30.12 + 42.55 = 73.92 \text{ MPa}$$

The outcome of the stress loss calculations is summarized in Table 8.4. The loss due to elastic shortening and creep are effectively zero. For the problem at hand, at the locations considered the net strain of concrete under self-weight is tensile. Strictly speaking, there will be an increase in the tendon stress, as opposed to decrease. However, in practice, the change is assumed to be zero. This is the common case, since at most design-critical sections along a member, tendons are located in the tensile zone of the section.

Table 8.4 Summary of long-term stress losses at second support

	MPa	%
Elastic shortening (ES)	0	0
Creep (CR)	1.25	2
Shrinkage (SH)	30.12	41
Relaxation (RE)	42.55	57
Total	73.88	100

NOTE

1. Reduced for day 3

REFERENCES

Aalami, B. O. (2014), *Post-Tensioned Buildings; Design and Construction*, PT Structures Inc., Palo Alto, www.PTStructures.com, 450 pp.

PCI (2017), *Deign Hand Book*, Precast/Prestressed Concrete Institute, Chicago, IL, 8th ed.

Zia, P., Preston, K., Scott, N., and Workman, E. (1979), *Estimating Prestress Losses*, ACI Committee 423 Report, Concrete International, June, pp. 32–38.

Chapter 9

Tendon layout and detailing

9.1 DISTINGUISHING FEATURES IN DETAILING PT AND RC SLABS

Post-tensioning adds precompression to floor slabs. This is the primary factor that differentiates the arrangement of reinforcement between conventionally reinforced and post-tensioned floors.

9.1.1 Crack control and disposition of tendons

A. Distance of probable crack to next tendon: Figure 9.1 is a partial plan of a post-tensioned floor slab. It shows the distribution of axial compression in the slab from the tendons anchored at the slab edges. The precompression at a point located away from the slab edges does not strictly depend on the distance of that point to the next tendon. Rather, it is the function of how the force applied at the slab edge disperses in the slab away from the anchors. For the configuration shown in the figure, the points at the center of the slab in parts (a) and (b) have the same precompression, but are distanced differently from the next tendon.

The contribution of unbonded post-tensioning to crack control at a point is primarily through the precompression it provides – not the distance to the next tendon. This is not the case for conventionally reinforced floors. In conventionally reinforced floors, non-prestressed reinforcement is commonly provided where there is potential of cracking from local tensile stresses. For crack control, the reinforcement is generally closely spaced – typically 1.5 slab thickness, or 500 mm.

Bonded tendons, in addition to precompression, contribute to crack control through their bond with concrete similarly to non-prestressed reinforcement.

The preceding observation allows the tendons to be spaced widely apart without losing their mitigating impact from precompression on crack control at locations away from the anchors.

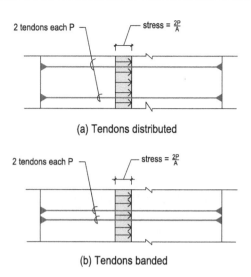

Figure 9.1 Partial plan; precompression in slab from PT tendons.

B. **Precompression and concentrated load:** For the design of floor slabs, in addition to uniformly distributed loads, major building codes require the slab to be checked for a specified concentrated load at any location on the slab.

Concentrated loads can result in tensile stresses at slab soffit below the point of application of the load. In RC (conventionally reinforced) floors, closely spaced bottom reinforcement – often in the form of a mesh – resist the tensile stresses from concentrated loads. Figure 9.2 shows a section through a slab that highlights the effects of precompression and load application.

Part (b) of the figure shows the precompression in the member from post-tensioning. Under no load, the precompression extends through the depth of the member. The application of load (part c) results in local redistribution of precompression.

The presence of precompression provides the member with the capacity to resist some transverse load, before tensile stresses develop. For this reason, in PT floors bottom reinforcement for crack control under moderate local load may be eliminated or not closely spaced.

9.1.2 Development of floor strength

Figure 9.3 illustrates the mode of failure of column-supported flat slabs under increasing uniform load. This is on the assumption of adequate

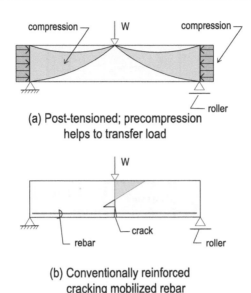

Figure 9.2 Section through slab showing precompression and load application.

punching shear strength. At failure the slab folds down to collapse between the line of supports in one of the principal directions.

In either of the two failure modes, the entire reinforcement that crosses the hinge lines shown will be activated. The activated reinforcement fully develops its yield strength and contributes to the resistance of the slab.

The location of reinforcement within the tributary of a column support is immaterial for the strength of the panel that the column serves. The failure mode hinge lines extend through the entire width of the tributary from support to support, thus mobilizing all the reinforcement within the tributary.

This observation supports the flexibility in position of reinforcement for safety requirements. As long as the total required reinforcement of each design strip is placed within the same tributary width, the reinforcement makes full contribution in avoiding collapse in the configuration shown in the figure.

For crack mitigation in service condition, the presence of bonded reinforcement is required where high tensile stresses develop. This is at or near the column supports. For this reason, in RC floors, and for service condition only, the concentration of reinforcement near the column region is beneficial. This is the underlying reason for the arrangement of reinforcement in RC floors according to column-strip/middle-strip layout.

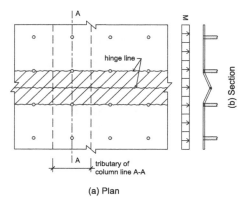

Slab Failure by Hinge Formation

Figure 9.3 Failure mechanism modes of column-supported floors.

For post-tensioned floors reinforced with unbonded tendons, the contribution to crack control comes from precompression. The precompression is not the function of distance to the next tendon. Compression is generally applied at the edge of the slab and disperses throughout the floor. For this reason, ACI 318 does not require the distribution of tendons to be based on column-strip/middle-strip arrangement.

9.2 TENDON ARRANGEMENT

From the preceding explanation, for a post-tensioned column-supported floor the arrangement of reinforcement within a design strip is not critical for safety considerations, so long as each design strip includes the total reinforcement that is required to prevent its collapse.

9.2.1 Tests on tendon arrangements

Load tests on post-tensioned column-supported floors conducted at the University of Texas [Burns, et al., 1971; Smith et al., 1974] used essentially the same amount of post-tensioning in each test model, but different tendon arrangements. Figure 9.4 shows the plan view of the different tendon arrangements considered. The tests' objective was to investigate the arrangement of tendons on load carrying capacity of post-tensioned column-supported floors.

The tests included distributed-distributed to grouped-grouped arrangement of tendons. The tests concluded that the tendon arrangement options shown in the figure all develop the required design strength for uniform loading. This led to the conclusion that from the tendon arrangement

Tendon layout and detailing 199

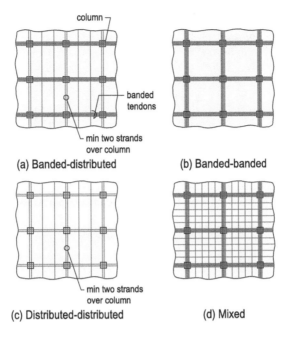

Figure 9.4 Tendon arrangement options.

options shown, the selection of tendon layout may be based on considerations other than the impact of arrangement on floor strength.

9.2.2 Tendon arrangements in practice

Depending on the geographical location, the construction practice and whether the post-tensioning system is bonded or unbonded, widely different tendon arrangements are practiced. All can be designed to meet the serviceability and safety stipulations of EC2 and ACI 318 building codes. The following illustrates several of the options.

9.2.2.1 Banded-distributed layout

The option is primarily practiced in the US. The following are several examples from the US and other regions.

A. **Orthogonal arrangement of supports**: Where supports are arranged on an orthogonal grid the banded and distributed tendons will be normal to one another. Figure 9.5 shows a typical arrangement of banded-distributed tendon layout for unbonded construction. Figure 9.6 is the same style of tendon arrangement using the bonded system and

Figure 9.5 Banded-distributed view of unbonded tendon layout; USA.

Figure 9.6 Banded-distributed view of bonded tendon layout; USA.

Figure 9.7 Banded-distributed tendon layout using bonded tendons; Jordan.

plastic ducts. An example of banded-distributed tendon layout using the bonded system with metal ducts is shown in Figure 9.7.

B. Irregular support layout: Where the supports are not on an orthogonal grid, in one direction tendons are grouped and swerve over the

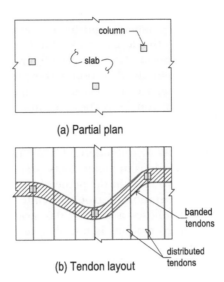

Figure 9.8 Tendon layout for non-orthogonal arrangement of supports.

Figure 9.9 Unbonded tendon layout for non-orthogonal arrangement of supports; Brazil.

line of non-aligned supports. Figure 9.8 shows a partial plan of a slab with non-aligned support layout, where the grouped tendons follow the line of non-aligned columns. In the orthogonal direction tendons are distributed and placed parallel to one another as shown in part (b) of the figure.

Figure 9.9 shows an example of grouped tendons swerving over non-aligned supports. In the other direction tendons are parallel to one another.

For non-orthogonal arrangement of columns, at the design stage, the design strips are likely to be non-orthogonal to one another. Each design strip is designed for the post-tensioning and reinforcement based on its actual geometry and force demand. Figure 9.10, parts (a) and (b), illustrate the design strips of a floor with non-orthogonal arrangement of supports. Part (c) of the figure shows the tendon layout at construction. In one direction (left–right in the figure) the tendons from the analysis are grouped and arranged over the non-aligned supports. In the orthogonal direction, tendons are distributed parallel to one another.

The arrangement shown is preferred for ease of installation. It affords avoiding weaving of tendons. The arrangement does not work well with bonded tendons in metal ducts, since the ducts do not have adequate flexibility to bend in the plane of the slab.

C. **Selection of the orientation of grouped tendons**: Where the arrangement of supports follows a rectangular grid, the selection of the direction for grouped tendons is not obvious. Figure 9.11 shows the preferred guideline for tendon arrangement in the absence of specific features of the support arrangement that may favor grouping the tendons in a given direction.

Arrangement of tendons shown in part (c) of the figure results in less wedge-shaped regions along the slab perimeter, where low precompression from prestressing may require addition of non-prestressed reinforcement for crack control.

9.2.2.2 Distributed-distributed layout

In this style, tendons in both directions are arranged at essentially equal spacing and parallel to one another. Figures 9.12 and 9.13 show two examples of the arrangement, one is using unbonded and the other bonded tendons. The arrangement is more common for bonded tendons, since the tendon ducts are not flexible enough to swerve in following the line of non-aligned columns.

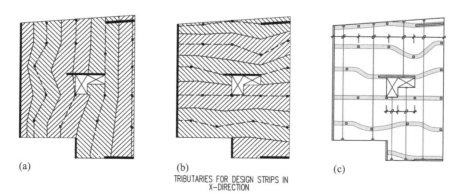

(a)   (b) TRIBUTARIES FOR DESIGN STRIPS IN X-DIRECTION   (c)

Figure 9.10 Plan; design strips and tendon layout.

Tendon layout and detailing 203

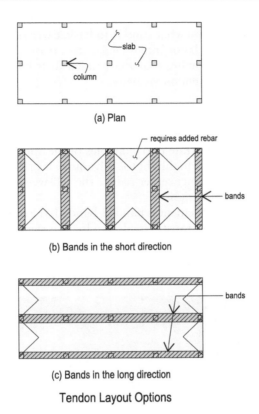

Figure 9.11 Options for arrangement of grouped tendons.

Figure 9.12 Distributed-distributed arrangement of unbonded tendons; Brazil.

The arrangement of tendons in distributed-distributed style requires sequencing of tendons somewhat similar to basket weaving. Both the planning and layout are more labor-intensive. For this reason, it is not generally the preferred option, knowing that the banded-distributed tendon layout can be installed without tendon weaving.

9.2.2.3 Banded-banded layout

Banded-banded tendon arrangement is advantageous for several reasons. Figures 9.14 and 9.15 illustrate two examples of the application of banded-banded tendon layout that highlight the advantage it offers. Voided spheres or waffles placed at the panel center reduce the self-weight of the slab leading to more economical designs.

When the tendon arrangement is banded-banded, the interior of the panel must be detailed to safely transfer its design load to the panel supports.

ACI 318-19 does not recommend the banded-banded distribution of tendons at this time. But, due to its inherent advantages and its validity based on mechanics of solids and tests, the arrangement is practiced.

Figure 9.13 Distributed-distributed arrangement of bonded tendons; Dubai.

Figure 9.14 Banded-banded tendon layout using voided spheres; Latvia.

Tendon layout and detailing 205

Figure 9.15 Banded-banded tendon layout using waffles; Fortaleza.

9.2.2.4 Irregular tendon layout

In theory, any tendon arrangement can be permissible, if the presence and contribution of the tendons are recognized and accounted for in the design satisfying the mechanics of solids.

Possibly, for this reason, EC2 does not prescribe specifics of tendon arrangements. Rather, focusing on compliance with serviceability and safety, it leaves it to the design engineer to decide on the arrangement of tendons for specific conditions.

9.3 TENDON PROFILE

9.3.1 Common conditions

A. **Interior span**: Traditionally, tendon shapes were selected to follow parabolic curves. The practice was based on the approximation of the lateral force exerted by the tendon to its concrete container as uniform force. Substitution of the tendon by uniformly distributed force facilitated the member design for hand calculation. Figure 9.16 shows the typical condition of an interior span, the tendon profile and the modeling of the tendon force. The tendon shown in the figure is made up of four parabolic curves each resulting in a uniformly distributed force.

The selection of a tendon profile in the shape of a parabola requires chairs, or other means along the tendon length on site, to make sure that the profile assumed in design is replicated in construction. Typically, tendon profile control chairs are provided at 1 m intervals. Other options are also available.

Tendons can provide greater uplift to counteract the gravity loads if they are profiled as shown in Figure 9.17 – referred to as bath tub profile. Parts (b) and (c) of the figure show the forces the tendon exerts to the member (balanced loads) in addition to precompression.

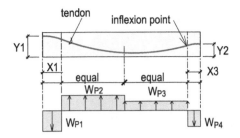

(a) Reversed parabola with two inflexion points

Example for Tendon Profiles and Balanced Loads

Figure 9.16 Typical tendon profile and modeling of its lateral force for common condition.

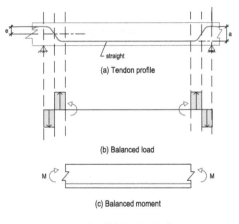

Bath Tub Tendon Profile

Figure 9.17 Bath tub tendon profile provides greater uplift than reversed parabola.

From a construction standpoint, the effort in maintaining the design-prescribed tendon profile can be reduced if less chairs are used to maintain the profile. In theory, tendon profile under self-weight as shown in Figure 9.18 provides a larger uplift than the commonly used reversed parabola. This is under the assumption that the tendon span exceeds the minimum length [Wicke, M., 2005]. In this case the number of control points can be reduced. This results in saving of labor and material.

B. Cantilever: Cantilever tendons are mostly designed to control deflection, and to counteract moments from gravity loads at the cantilever

Tendon layout and detailing 207

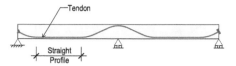

Figure 9.18 Tendon profile under self-weight.

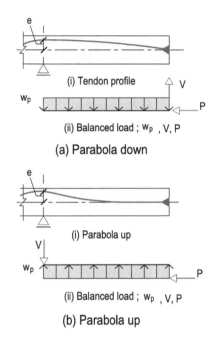

Figure 9.19 Cantilever up and cantilever down tendon profile options.

support. For moment control, the contribution of the tendon is the fixed value of tendon force (P) times the eccentricity of the tendon at the face of the cantilever support (e). In other words, the contribution of the tendon to moment resistance (M) at the face of support is independent of the tendon profile in the cantilever.

$$M = Pe \tag{9.1}$$

Figure 9.19 shows the parabola up profile in a cantilever span. The profile shown provides the same moment at the face of support. But parabola down

208 Post-tensioning in building construction

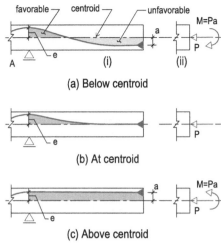

Figure 9.20 Favorable and unfavorable moment regions for upward deflection of the cantilever.

(part a) contributes more to the upward deflection at the tip of the cantilever – hence it is the preferred option for deflection control.

For tendon contribution to upward deflection, the profile that gives the largest moment of the post-tensioning moment diagram about the cantilever support is more effective. Figure 9.20 illustrates three options for tendon layout in a cantilever span. The options shown all have the same moment at the face of support, but the profile in part (c) of the figure provides the larger upward deflection.

9.3.2 Tendon at discontinuities

A. **Ramps**: Ramps in parking structures and steps in floor slabs are the common conditions of discontinuities in flat floor slabs. In the absence of design software that can model a ramp as it occurs, one option to obtain the design values is to model for design the ramp span with no kink. Figure 9.21 shows an option for the structural model of a ramp span for the calculation of its design actions.

The analysis model uses a straight span with the same thickness as the prototype and the specified load of the ramp (part b of the figure). The eccentricity (*e*) of the tendon obtained from the design of the span design model is noted. The tendon in the prototype is laid out

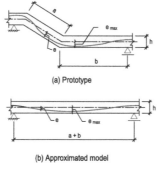

Figure 9.21 Structural approximation of slab at angle change.

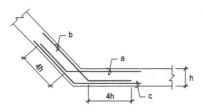

Figure 9.22 Detailing of ramp kink.

following the eccentricity determined from the analysis model. The lateral force exerted by the tendon to the ramp will be the same as the analysis model.

There will be local in-plane forces at the kink in the ramp. These forces are in self-equilibrium. The impact of the concentrated forces will not be of design significance beyond approximately two times the slab thickness from the kink. Figure 9.22 shows the added reinforcement in the ramp that accounts for the local forces at the kink.

B. **Steps in slabs**: Figure 9.23 shows a step in the slab along with its treatment for analysis and construction. The drop can be modeled as a uniform slab for analysis. For construction, the step is ideally detailed as shown in part (b) of the figure. The slab overlap affords smooth transition of the tendon through the drop.

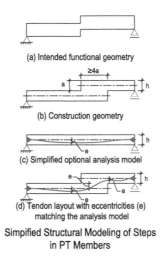

Figure 9.23 Analysis and detailing option for drop in slab elevation.

If the analysis tool used for design does not have the capability to model the location as shown in part (b) of the figure, the alternative modeling shown in part (c) can be used for analysis. For construction, the location is detailed as shown in part (d) of the figure. The tendon profile specified through the drop is detailed with the same eccentricity arrived at from the analysis model shown in part (c).

The location must be adequately detailed with non-prestressed reinforcement not shown in the figure.

C. **Tendon along wall**: Along and over the wall supports tendons are kept at high point and straight. Figures 9.24 and 9.25 show the schematics and construction example of the tendon arrangement over a wall support. At the face of the wall, moments are likely to cause tension at the top of the slab. The tendon at the slap top reduces the tension.

Along the end walls and edge beams the tendon may be straight, but lowered to the slab edge to be anchored at the centroid of the first span (parts b and c of Figure 9.24).

9.4 NON-PRESTRESSED REINFORCEMENT

In addition to post-tensioning, all post-tensioned floors include some non-prestressed reinforcement not specified in building code serviceability and safety requirements.

9.4.1 Trim bars

Figure 9.26 shows a flow chart that identifies the different categories of the non-prestressed reinforcement – referred to as trim bars – that are used in

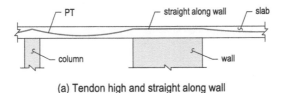
(a) Tendon high and straight along wall

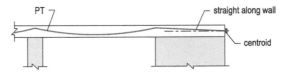
(b) Tendon straight over wall

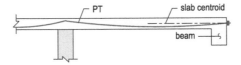
(c) Tendon anchored at slab edge

Tendon Profile along Wall

Figure 9.24 Tendon profile over and adjacent to wall.

Figure 9.25 Tendon is placed at top and straight over the wall.

post-tensioned floor slabs. The objective of the bars is to improve the in-service performance of the slab. Trim bars are identified as group 5 in the following flow chart. Referring to the flow chart numbers, the categories are:

1–Bars required by analysis to meet the code-specified requirements for the serviceability and safety of the structure. These are described in Chapter 4 according to the building code used for design.

212 Post-tensioning in building construction

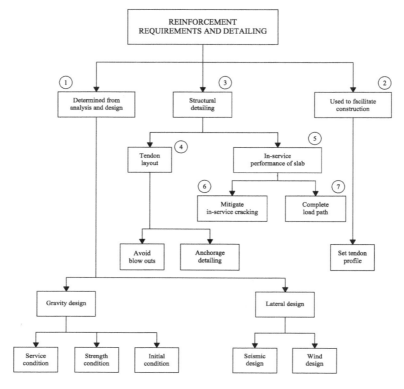

Figure 9.26 Categorization of non-prestressed detailing reinforcement.

2– Bars used to facilitate tendon layout and other construction measures. These are handled by the post-tensioning supplier to meet the specifics of the hardware used.

3– Structural detailing. The bars are subdivided into two categories.

4– Bars required to contain the tendons in their installed position and avoid concrete rupture arising from high localized tendon force. The common conditions are typically shown on the construction drawings of the structural engineers for the post-tensioning system envisaged.

5– Bars used to improve the in-service performance of the slab. These fall into two categories.

6– Bars used to address local concentration of stress that can lead to crack formation. These are typically bars at locations of abrupt change in the geometry of the slab. The common conditions of these bars are listed and commented on below.

7– Bars to complete load path. Floor slab design is commonly based on breaking the floor into design strips. This is followed by determination of the total amount of reinforcement required for each design

Tendon layout and detailing 213

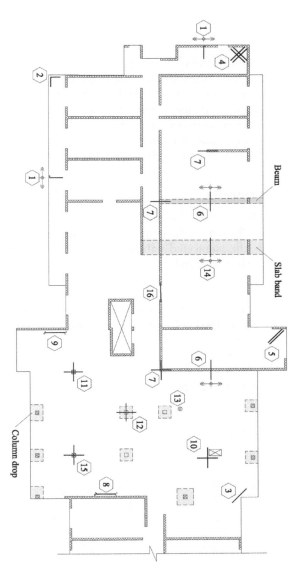

Figure 9.27 Partial plan of a post-tensioned floor identifying the trim bar categories.

strip. Subsequent to the design, the structural engineer reviews the outcome to make sure that non-prestressed reinforcement is available or added where concentrated loads act.

Figure 9.27 is the partial plan of a post-tensioned floor that identifies the different non-prestressed reinforcement bars commonly checked for improved service performance of the slab. The bar details are commonly shown on construction drawings and referred to on plans as minimum reinforcement. The numbers in the figure refer to the details that are shown on the following pages. They apply to similar situations.

In the following section, the trim bars are listed on the premise that there is neither a bottom nor top mesh specified for the floor. In the US it is not common to provide bottom mesh reinforcement. However, provision of bottom mesh is practiced in many parts of the world.

The following lists the categorization of the bars shown on the plan. The list is followed by an example of each detail. (Figures 9.28 to 9.43)

9.4.2 Trim bar details

The reinforcement listed and shown in the details is the 'minimum' amount. If the reinforcement shown on the construction documents from other considerations exceeds the suggested values in the following, none is added. Otherwise, the shortfall is added.

Detail 1
Reinforcement along the edge of slab, both where the slab edge is supported or is free. In the US L-bars are common. U-bars are used where EC2 is dominant.

Detail 2
Reinforcement at slab corners.

Detail 3
Reinforcement at re-entrant corners.

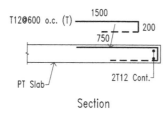

Figure 9.28 Detail 1.

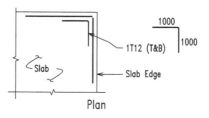

Figure 9.29 Detail 2.

Detail 4
Reinforcement at slab corners that are supported on both sides. It applies to all levels of a structure at corners that are supported by walls on both sides.

Detail 5
Reinforcement to mitigate cracking at slab corners that are tied to perimeter walls. This reinforcement is recommended for the first two to three levels of tower structures, whereas reinforcement shown in detail 4 applies to all levels of a building. Moreover, detail 5 applies to conditions, where a slab edge is supported on either one or two sides, whereas detail 4 applies only where an exterior slab corner is supported on both sides.

Detail 6
Distributed top reinforcement transverse to beams and walls. The trim bar extends over the entire length of the member.

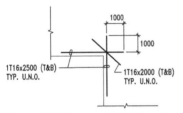

Minimum Reinforcement at Re-entrant Corners

Figure 9.30 Detail 3.

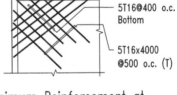

Minimum Reinforcement at Supported Slab Edges

Figure 9.31 Detail 4.

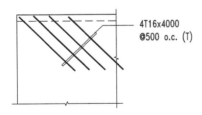

Minimum Crack Control Bars at Supported Slab Edges

Figure 9.32 Detail 5.

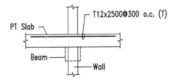

Minimum Top Reinforcement Transverse to Interior Walls and/or Beams

Figure 9.33 Detail 6.

Detail 7
Top reinforcement at the tip of a wall or beam support.

Detail 8
Reinforcement along interior walls that are tied to slab. This reinforcement is generally required for the first two to three levels of a building tower.

Detail 9
Reinforcement along exterior walls that are tied to slab. This reinforcement is generally required for the first two to three levels of a building tower.

Detail 10
Reinforcement around small openings that are not handled in design. Larger openings are generally modeled in the analysis and are designed for. The detail covers openings up to 1 m in size. Reinforcement for larger openings is to be designed for.

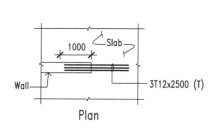

Figure 9.34 Detail 7.

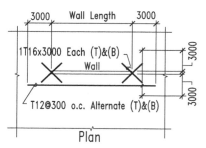

Figure 9.35 Detail 8.

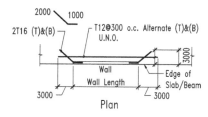

Figure 9.36 Detail 9.

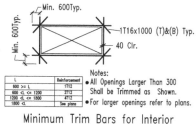

Figure 9.37 Detail 10.

Detail 11
Bottom reinforcement over column supports.

Detail 12
Reinforcement in column drops/drop panels.

Detail 13
Stirrup for openings near and within the vicinity of critical punching shear perimeter of a column, where the presence of opening has not been accounted for in design. The stirrups are intended to compensate for the loss of strength from opening in slab.

Detail 14
Reinforcement in slab bands.

Detail 15
Reinforcement below concentrated loads. At design time, concentrated loads may have been assumed smeared over the width of the design

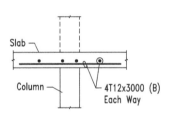

Bottom Reinforcement at Column Support

Figure 9.38 Detail 11.

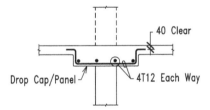

Minimum Reinforcement for Drop Cap/Panel

Figure 9.39 Detail 12.

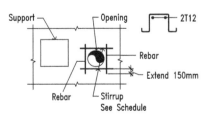

Stirrup for Opening Next to Support

Figure 9.40 Detail 13.

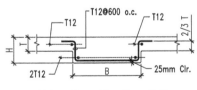

Minimum Slab Band Reinforcement

Figure 9.41 Detail 14.

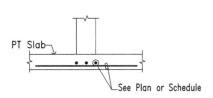

Reinforcement Bellow
Concentrated Load

Figure 9.42 Detail 15.

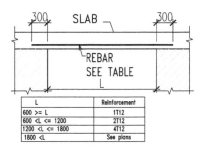

Minimum Bottom Bars Over
Gap in Continuous Support

Figure 9.43 Detail 16.

strip they act on. The objective of the reinforcement is to spread the load over the design section assumed for its analysis and design.

Detail 16
Where there is interruption in wall support without a header, the unsupported length of the slab is provided with bottom rebar shown in detail.

OTHER CONDITIONS

There are other conditions not listed. Holdowns at the end of shear walls that terminate on the slab apply up-and-down forces. The required reinforcement detail depends on the intensity of the force. The embedment in the slab for the holdowns is generally shown on the structural drawings.

REFERENCES

Burns, N. H. and Gerber, L. L. (1971), "Ultimate Strength Test of Post-Tensioned Flat Plates," *Journal of the Prestressed Concrete Institute*, V. 16, No. 6, November–December 1971, pp. 40–58.

Smith, S. W., and Burns, N. H., (1974), "Post-Tensioned Flat Plate to Column Connection Behavior," *Journal of the Prestressed Concrete Institute*, V. 19, No. 3, June–May 1974, pp. 74–91.

Wicke, M. (2005), "Spannglieder ohne Verbund im Hochbau – Bemessung und Ausfuerung," Expertenforum Beton 2005, Vorgespante Flachdecken, Zement & Beton, Vienna, April 2005, pp. 8–13.

Chapter 10

Post-tensioning construction in buildings

10.1 POST-TENSIONING IN BUILDING CONSTRUCTION

Extended development and application of post-tensioning in building construction started in the mid-1950s. Today, post-tensioning is common in building construction. Chapter 1 covers the historical development of its application.

Two post-tensioning hardware types are mostly used in building construction. These are the bonded and unbonded systems. They differ in the protection they provide to the environmental elements, and the manner in which they respond to applied load. The characteristics of each hardware style and the response of each to applied load is explained in Chapter 3.

The focus of the following is a brief review of the two system components and their application in construction.

10.2 SYSTEM COMPONENTS

Typical post-tensioning systems used in building construction consist of:

(i) Prestressing steel; prestressing strand.
(ii) Prestressing tendons.
(iii) Dead and live end hardware, including the reinforcement for dispersion of high local stresses at tendon anchorage.

10.3 PRESTRESSING STEEL AND STRAND

A. **Prestressing steel**: High strength steel is used for prestressing application. Most steel used has ultimate strength of 1,860 MPa. Higher strength steel is also available. Figure 10.1 shows the typical mechanical characteristics of the prestressing steel used in common construction.

220 Post-tensioning in building construction

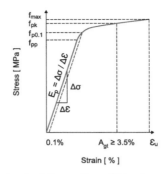

Figure 10.1 Mechanical properties of prestressing steel.

Figure 10.2 Cross-section of a typical strand.

The primary material parameters of the common prestressing steel marked in Figure 10.1 are:

f_{pk} = 1,860 MPa

$f_{p0.1}$ approximately 1,640 MPa
$f_{p0.1}/f_{pk}$ approximately 0.88

B. Prestressing strand

Practically all prestressing steel used in common building construction comes in the form of stands. Figure 10.2 shows the cross-section of typical seven-wire prestressing strands.

- The common strand diameters used in building construction are 12.5 mm to 15.7 mm.
- The cross-sectional area of the strands commonly used is 93 mm^2 to 150 mm^2.
- The steel grade (ultimate strength) for common strands is in the range of 1,770 MPa to 1,860 MPa.

Other strands with fewer number of wires and different characteristic values are also available, but less common.

Smaller diameter strands (12.5 mm) are typically used in cast-in-place floor slabs. Larger diameter strands (150 mm) are more common in precast elements and bridge construction.

10.4 PRESTRESSING SYSTEMS

The characteristics and response to load of the commonly used bonded and unbonded systems are described in Chapter 3. The following explains the construction features of each.

10.4.1 Bonded system

Bonded systems typically consist of tendons that house up to five strands in a metal sheet or plastic duct. Figure 10.3 illustrates the components of a typical grouted tendon and the nomenclature of its parts. Figure 10.4 is the view of the stressing end of a typical grouted tendon with flat duct housing four strands. At installation the tendon ends are provided with added local reinforcement to disperse the concentrated forces from the tendon.

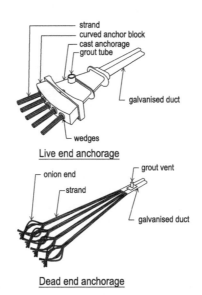

Figure 10.3 Components and nomenclature of typical dead and live ends of bonded tendon.

Figure 10.4 View of a typical bonded tendon stressing end assembly (courtesy BBR).

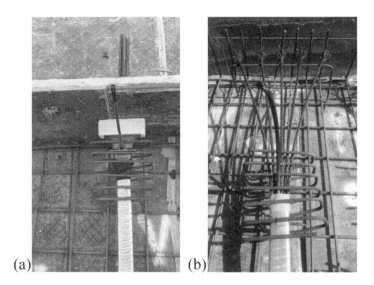

Figure 10.5 Live and dead end details of a bonded post-tensioning system ready to receive.

The amount and the configuration of the dispersion steel at the tendon ends depend on the number of strands in the tendon and the configuration of the stressing block. Post-tensioning suppliers generally provide the details of the dispersion steel required for their hardware.

At each tendon end and at several intervals at high point locations along the long tendons vent tubes are provided. The tubes allow entry of grout into the duct or discharge of air from the duct displaced by the grout.

Figure 10.5 shows the live and dead ends of a bonded system in place ready to receive concrete. The anti-bursting steel for dispersion of local force and the vent for injection of grout or exit of air are secured in place before concrete is cast. At the live end the polystyrene block provides the recess at the slab edge for stressing the strands.

Where the large size of the anchorage assembly interferes with the reinforcement in the slab; or the stressing end does not extend to the member edge; or the congestion of reinforcement, such as presence of wall at the member end does not allow for the stressing end to extend to the slab edge, stressing pans at the top of the slab provide the stressing option. Figure 10.6 shows a stressing pan at the top of a slab.

Stressing of strands at the slab surface through the pan requires jacks with a curved nose extension.

10.4.2 Unbonded system

The hardware of unbonded systems is typically smaller in size, since each tendon holds only one strand. The smaller force at the live and dead end compared to the typical bonded system affords smaller anchorage casting and less anti-splitting reinforcement. Figure 10.7 is an example of the live end and Figure 10.8 an example of the dead end in the typical application of unbonded tendons in place ready to receive concrete.

Figure 10.6 Stressing pan at slab top.

Figure 10.7 Example of an unbonded stressing end in place.

224 Post-tensioning in building construction

Figure 10.8 Example of an unbonded dead end in place ready to receive concrete.

Typically, for the slab edge it is sufficient to provide two 12 mm bars behind the anchor piece normal to the tendon. Figures 10.9 and 10.10 are the ACI 318-19 recommended details for bursting steel. Figure 10.10 applies to detailing of grouped tendons at stressing ends. The details apply to both 13 mm and 15 mm strands.

Figure 10.11 is the construction view of unbonded grouped tendons anchored at the slab edge with anti-bursting reinforcement in place.

10.5 CONSTRUCTION

Material and labor costs are two of the major considerations for the selection of a construction scheme. For post-tensioned floor systems the material quantities for specific applications are somewhat well established. The global variations in unit cost of the material and labor largely determine the selection of the post-tensioning option compared to other alternatives.

10.5.1 Quantities

The post-tensioning and rebar quantities depend on the configuration of the floor slab, its support layout, specified load and concrete strength. In addition they vary depending on the perception and practice of local engineers regarding the concepts and behavior of post-tensioned members.

Typical quantity values for common residential and commercial post-tensioned floors satisfying EC2 and ACI 318-19 code requirements are:

Bonded post-tensioning system

Specified load: 1 kN/m^2 superimposed dead load; 2.5 kN/m^2 live load

- PT: 3-4 kg/m^2
- Rebar: 5 kg/m^2

Post-tensioning construction in buildings 225

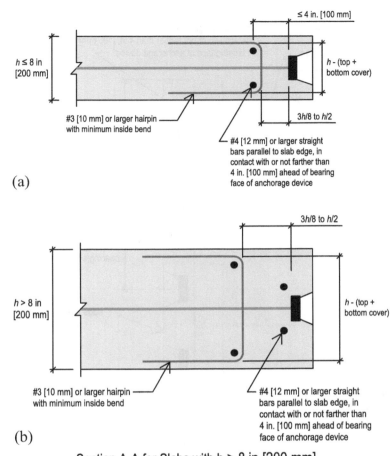

Figure 10.9 Recommended anti-bursting reinforcement (ACI 318-19).

Specified load: 3 kN/m² superimposed dead load; 3 kN/m² live load

- PT: 3.5-5 kg/m²
- Rebar: 7-9 kg/m²

Unbonded post-tensioning system
   Specified load: 1 kN/m² superimposed dead load; 2.5 kN/m² live load

- PT: 3.75 kg/m²
- Rebar: 6 kg/m²

Local practice of adding mesh and non-building code specific reinforcement results in higher values.

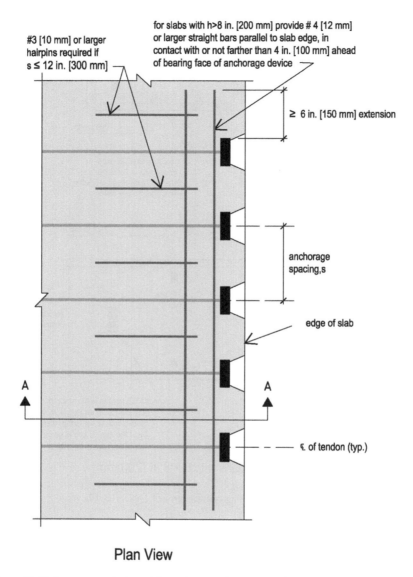

Figure 10.10 Recommended anti-bursting reinforcement for grouped tendons (ACI 318-19).

10.5.2 Construction cost

Construction cost varies greatly depending on the country. The central items of labor, concrete, tendon and rebar are subject to large local variations. Depending on the country, the cost ratio of in-place unit price of 1 kg of post-tensioning to 1 kg of rebar varies between 1 and 7.[1]

Post-tensioning construction in buildings 227

Figure 10.11 Unbonded tendon group at stressing end with anti-bursting reinforcement in place.

Where labor is expensive, it becomes more economical to use post-tensioning. For example, in California where bottom mesh is not commonly used and labor is expensive, the installed cost of material and labor of 1 kg of rebar is essentially the same as 1 kg of post-tensioning. Since the total weight of reinforcement using post-tensioning is less than that of using conventionally reinforced floors, from a cost standpoint, the post-tensioning alternative becomes the preferred option.

The primary steps of typical post-tensioned floors in building construction are:

- Forming
- Rebar installation
- Tendon installation
- Casting of concrete
- Stressing and finishing

10.5.3 Local practice

A. Arrangement of tendons

Chapter 3 concludes that the arrangement of tendons in a floor slab is not critical to the load-carrying capacity of a post-tensioned floor as long as each design strip is provided with the design-required amount of reinforcement. The arrangement can be made following the construction preference, with due consideration to detailing.

Options for arrangement of tendons detailed in Chapter 3 are:

(i) Grouped in one direction and distributed in the orthogonal direction.
(ii) Distributed in both directions.
(iii) Grouped in both directions.

For solid flat slab construction, grouped in one direction and distributed in the orthogonal direction results in least labor. The tendon layout can be sequenced to avoid weaving at tendon intersections. Figure 10.12 shows an example of grouped-distributed tendon layout.

Other tendon arrangement examples, such as voided spheres and waffles are given in Chapter 9.

B. Mesh reinforcement

Addition of bottom mesh is based on local practice, as opposed to design or code requirement. The application is primarily transition of practice of conventionally reinforced concrete to post-tensioned alternative. In the US bottom mesh is not generally provided, unless required by design. In many other locations provision of bottom mesh is common practice. Where used, it is generally 10 or 12 mm bars spaced at 200 to 300 mm spacing.

Top mesh is not common. Some jurisdictions require top mesh if slab thickness exceeds 300 mm.

Figure 10.13 shows examples of an unbonded and bonded slab construction without bottom mesh ready to receive concrete.

Figure 10.14 shows examples of unbonded and bonded post-tensioned floors with bottom mesh ready to receive concrete.

Figure 10.12 View of grouped-distributed tendon layout.

Post-tensioning construction in buildings 229

(a) Two-way slab reinforced with unbonded tendons   (b) Two-way slab reinforced with bonded tendons

Figure 10.13 Examples of floor slabs without bottom mesh ready to receive concrete.

(a) Unbonded tendon layout with bottom mesh   (b) Bonded tendon layout with bottom mesh

Figure 10.14 Examples of floor slabs with bottom mesh ready to receive.

C. Securing tendon profile

The shape of tendon length in place – referred to as tendon profile – is determined by design. It is the tendon shape and the force in the tendon that provides the up-and-down forces configured by design to counteract the gravity loads to improve slab performance. The lateral force exerted by a tendon to the concrete that contains it is in the plane of the tendon profile.

In placing tendons on site it is critical to note the following:

- Tendons must follow as closely as practical the profile specified on construction drawings.
- Tendons must be secured in position on the forms such that during the installation of subsequent reinforcement and placing of concrete they retain their position.
- At the live and dead ends, where tendon force is transferred to concrete, adequate reinforcement must be provided to disperse the transfer of concentrated tendon force.

- At locations of sharp tendon curvature, specifically where tendons are near the slab surface, supplemental reinforcement must be provided to ensure that lateral tendon forces do not lead to local rupture of concrete.
- Depending on local practice, tendons are secured in position at about 1 to 1.5 m spacing.

Many contractors use individual chairs to set and maintain the tendon profile. Chairs are typically set at about 1 m spacing. The followng shows several options from practice in securing the tendon profile in place.

Figure 10.15 shows chairs with adjustable heights. The required tendon height marked on the tendon duct is achieved by pulling apart the legs of the U-chair at installation.

Figure 10.16 shows the common practice in the US. Tendons are tied to the typical 12 mm bars that in turn are secured on chairs. At low points, where tendon height is small, plastic strips (slab bolsters) are used.

Other options are used to set and secure the tendon profile, including adjustable chairs. Figure 10.17 is an example of a chair with adjustable height and its application in place.

10.5.4 Construction sequence and cycle

A. **Installation sequence**: Construction sequence of floor slabs varies depending on the local practice and overall schedule of project.

Where the same crew places rebar and post-tensioning, the installation is sequenced. The target of the placing sequence is that no item, whether prestressed or non-prestressed, will be placed below an existing reinforcement or tendon – there is no weaving or threading involved. Each piece of reinforcement is simply dropped where it belongs and tied in position. A crew of four is expected to start and finish 50 m^2 of floor slab per day. This includes both prestressed and non-prestressed. The reinforcement, where necessary, is shaped to facilitate placing. As an example, beam

Figure 10.15 Adjustable height individual chairs; UAE.

Post-tensioning construction in buildings 231

Figure 10.16 12 mm bars on single chairs used to control and secure tendon profiles; USA.

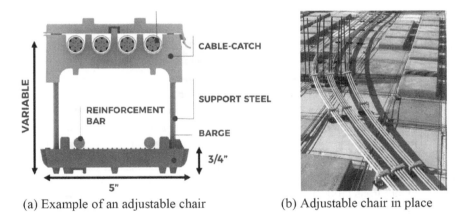

(a) Example of an adjustable chair (b) Adjustable chair in place

Figure 10.17 Adjustable tendon profile chair securing up to four unboned tendons in height (courtesy Impacto).

stirrups are provided with open tops. Where stirrups are required to be closed, the top segment is added afterwards as a separate piece.

Figure 10.18 illustrates the effort required by the prestressing crew to thread the tendons through the completed reinforcement cage of a beam. In part (a) of the figure, a tendon capped with a plastic bottle is pushed through the reinforcement case. After reaching the far end, in part (b) it is lifted and guided into the anchor casting at the beam end.

B. **Construction cycle**: A typical construction cycle is 5 days, starting on Monday and finishing with casting the floor on Friday. The weekend is used for curing of concrete cast on Friday. Faster cycles as short as one day have also been practiced, but are not common.

Figure 10.19 illustrates the typical 7-day cycle for a construction crew that installs both the prestressed and non-prestressed reinforcement concurrently.

(a) Tendon capped with a plastic bottle is pushed through the reinforcement cage

(b) Once reaching the end, tendon is led into the anchors installed at the beam end

Figure 10.18 Illustration of difficulty in tendon installation in beams with closed top ties.

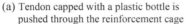

Activity	Day
	1 2 3 4 5 6 7 8 9 10
Construct Falsework	
Fix Bottom Reinforcement	
Fix Post-Tensioning Anchors	
Fix Post-Tensioning Tendons	
Fix Top Reinforcement	
Inspect	
Pour	
Initial Stress (25%)	
Final Stress (100%)	
Strip Formwork	
Grout Tendons (Not on critical path)	

Figure 10.19 Typical construction cycle of post-tensioned floor slabs.

10.6 STRESSING OPERATION

There are several considerations regarding the time and procedure of stressing tendons. The key in a successful stressing operation is to avoid rupture or spalling of concrete at stressing, and more importantly to make sure that the member shortens under the jacking force. The shortening of the member is essential for the contribution of post-tensioning to the member's bending strength. Chapter 7 provides details of shortening importance in member's strength.

10.6.1 Time of stressing

The understanding among many contractors is that the sooner the member is stressed, the better it compensates for shrinkage shortening. Early stressing helps to close early shrinkage cracks that may develop.

Early stressing means low modulus of elasticity of concrete. For floor slabs with multiple tendons, where tendons are stressed sequentially, this can result in larger stress loss in post-tensioning of previously stressed tendons from shortening of the member under jacking force of new tendons.

There is no common practice among post-tensioning professionals regarding the optimum time for stressing and the stressing sequence. The practice differs widely.

In the US tendons are stressed to full design force at one time. Typically, stressing is specified to take place when concrete reaches 20 MPa. Most hardware and splitting reinforcement is designed for stressing at about 15 MPa. Concrete is typically targeted to be cast on Friday. Tendons are stressed to full force on the following Monday.

In a number of locations, stressing is done in two stages. The following sequence is an example of a typical sequence.

- The first stressing takes place about 24 hours after concrete is cast and it has reached 25% of its specified strength, but not less than 12 MPa.
- The full stressing is done 2 to 3 days after concrete is cast and its strength has reached about 25 MPa.

At stressing, there are two considerations regarding the structural integrity of the region immediately behind the anchorage. Figure 10.20 illustrates the dispersion of the stressing force into the member and formation of splitting tension stresses.

First the compressive strength of concrete immediately behind the anchorage must be adequate to avoid crushing of concrete. Second, dispersion of compression from the smaller anchorage casting to the larger concrete section results in tensile stresses in concrete. Where required, tension bars normal to the tendon are used to resist the 'splitting tensile force.'

ACI 318-19 provides recommendations for detailing the anchorage location for the common strands in post-tensioned slabs. These are shown in

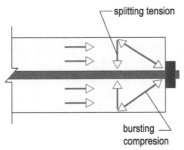

Dispersion of Force Behind Anchorage

Figure 10.20 Spreading of force at anchorage.

Figures 10.9 and 10.10. The recommended ACI detail applies to both 13 and 15 mm nominal diameter strands.

Stressing of tendons while concrete has not reached adequate strength, or if the anti-bursting reinforcement over the stressing region is not adequate can lead to spalling of concrete. Figure 10.21 is an example of concrete spalling at the stressing end of a group of tendons.

10.6.2 Stressing equipment

Post-tensioning in building construction uses mostly individually stressed strands – referred to as 'mono' strand construction. In mono-strand construction, each strand is individually engaged, pulled to the design force and seated. This is also true for bonded tendon ducts that house several strands.

In multi-strand construction all the strands of a tendon are pulled and seated simultaneously by a stressing jack that can engage more than one strand at a time. The number of strands being stressed simultaneously can be more than 60.

Two types of stressing equipment are common for single strand stressing. Figure 10.22 (a) is the kind that requires the strand tail to thread through the jack. Figure 10.22 (b) is a view of the jack that rides over the tendon tail extension. The latter enables tendons to be stressed and seated anywhere along the tendon length – not necessarily at the tendon end. This enables long tendons to be stressed at intermediate points, such as at construction joints. The former can be used only at the tendon end.

One option to provide continuity for stressing of strand segments is to use couplers between the stressing end of one tendon segment and the adjoining tendon segment intended to provide continuity of tendon force.

Recent developments in stressing and measurement of elongations include the combined stressing and elongation measurement equipment that automatically measures the force and the associated elongation. The

Figure 10.21 Spalling of concrete at stressing of grouped tendons.

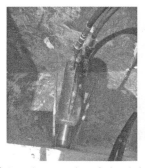 

(a) Stressing operation of a bonded tendon with strand threaded through the jack

(b) Stressing operation of an unbonded tendon, using a jack that slides over the prestressing tendon

Figure 10.22 Stressing jacks for mono-strand stressing.

Figure 10.23 Multi-strand jack in operation.

force-elongation measurement is automatically transferred to the stressing record file of the project [Oncrets, 2022].

Figure 10.23 shows a multi-strand jack used in heavy building construction, such as transfer beams. The stressing jack engages and pulls all the strands of the tendon simultaneously.

10.6.3 Elongation measurement

Measurement of strand elongation at stressing and its correlation with the design force of the strand is one of the central steps in the design of post-tensioned members. The specified elongation achieved on site is the testimony that the strand has been pulled and seated to the design force.

In most construction, the measurement of strand elongation is simply by steel tape to accuracy of typically 2 to 3 mm. Figure 10.24 shows the elongation measurement of a strand that has been pulled and seated.

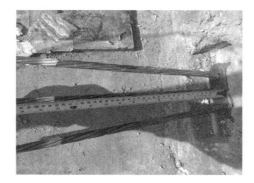

Figure 10.24 Elongation measurement in progress for a strand that has been pulled and seated.

Automated stressing and elongation measurement devices are also available. The force and elongation are measured and recorded electronically.

10.6.4 Evaluation of elongation

Chapter 8 explains that the elongation of a strand at stressing is directly related to the force in the strand. Attaining the design force for each strand is site-verified through the strand's elongation at stressing. The recording and correlating the computed and measured elongation are central steps of quality control of the construction process. For this reason, the elongation measurements and their recording are overseen by quality control personnel.

The site-measured tendon elongation measurements are matched against the computed values. Where discrepancy exceeds the variation expected in the construction process, the impact of the discrepancy on the design-intended performance of the post-tensioned member needs to be evaluated. Remedial measures may be needed if the amount and frequency of the discrepancy are evaluated to negatively impact the expected design performance of the member.

ACI 318 recognizes a 7% variation between the measured and computed elongation to be within the range of construction inaccuracies. It is impractical to achieve the required tolerance for all strands of a design member. The intent of the requirement is met if the aggregate of deviations for all strands of each design member is within the specified tolerance. A design member is viewed as a design strip of a floor slab or a beam in column-supported floors. For parallel beam and one-way slab construction the design member of the slab is viewed as a width along the beam equal to the distance between adjacent beams, namely a square panel of the slab between the beams.

A number of consultants allow for 10% deviation where bonded tendons and metal ducts are used.

10.6.5 Removal of shoring; propping

In common construction, shoring is removed after completion of stressing. Fully stressed floor panels are considered capable to carry their self-weight by combination of biaxial precompression and concrete strength reached at time of stressing.

Propping is required if the newly cast floor is subject to construction load or slab supports shoring of the floor level to be built immediately above it. In this case, three levels are typically shored to carry the weight of construction of one level above the last cast floor. Figure 10.25 shows the typical three levels of shoring for a multi-story construction.

10.7 GROUTING

Grouting is required for bonded construction. The void in a bonded tendon is commonly filled with cementitious grout. The grout has two important functions. First, the grout replaces the void in the tendon duct and cuts off air and moisture to the prestressing strand. Second, it provides an alkaline environment which inhibits corrosion.

From a structural standpoint the grout locks the deformation of the prestressing steel to the concrete that immediately surrounds it. This provides the deformation compatibility of bonded tendons with the encasing concrete. Deformation compatibility of bonded tendons enhances the load-resisting performance of the prestressing steel compared to the unbonded alternative system.

Grouting takes place soon after stressing is complete and stressing records are approved. Typically, this takes place 3 to 4 days after the

Figure 10.25 Shoring of several levels to support the construction of new floor.

Figure 10.26 Exposed stressing end of bonded tendon at slab edge being prepared for grouting.

member is stressed. It is not uncommon to delay the grouting of a specific member or floor region to coordinate with the construction schedule of other members.

The successful grouting is integral to the long-term durability of the post-tensioned system. In some jurisdictions the grouting crew need to go through specific training courses, pass the associated exam and be certified to perform grouting.

Tendons to be grouted are provided with a tube at one end to let the grout in. Additional tubes are provided at the far end and at selected high points along the tendon length to let the air out while grout is being pressure pumped. Once the tendon void is filled with grout, the outlets are sealed.

The grout mix is required to develop a minimum specified strength, typically exceeding 30 MPa. The grout is pumped with controlled pressure. Figure 10.26 (a) is the exposed view of the strands stressed, seated with the stressing pocket sealed for grouting (part b). Unlike the unbonded alternative where the stressing recess can be finished later, for grouted tendons the recess is sealed as shown in part (b) of the figure in order to avoid spilling of pressure-pumped grout.

Figure 10.27 shows the insertion of grout at one end and the ejection of grout with good consistency at the far end of a floor slab. The ejection of grout with good consistency at the far end signifies the successful completion of the operation when the grout vents can be sealed off.

10.8 FINISHING THE STRESSING RECESS

A. **Cutting the tendon tail**: The tendon tails protruding out of the slab edge necessary for jacking are trimmed back once the stressing is complete. The tendon tails are typically severed a minimum of 25 mm from the finished face of the member. The minimum cover to the member face is intended to satisfy the fire protection requirements.

(a) Technician feeding the grout at grout inlet

(b) Grout pouring out at the far end signals the duct having been successfully filled

Figure 10.27 Grouting operation of a bonded tendon.

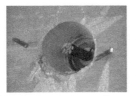

(a) Tendon tail trimmed off

(b) Grease and cap add corrosion protection

(c) Recess is covered with stiff grout flush with member surface

Figure 10.28 Finishing the stressing recess of an unbonded tendon.

There are several methods to cut the protruding tendon tail of unbonded tendons. These are flame cutting; use of rotary abrasive tool; special shear cut device; or electrical current. Flame cutting is the common practice in the US. The rotary abrasive tool is widely used elsewhere.

The tendon recess is cleaned from debris and loose material. The cut tendon tail is covered with a grease-filled cap. It is then pounded with stiff mortar mix to finish the recess in line with the member surface. Figure 10.28 shows the typical stages of finishing the stressing end of a single strand tendon.

10.9 MAINTENANCE

Well-executed post-tensioned construction is expected to be subject to the same maintenance routine as its conventionally reinforced alternative.

Post-tensioned tendons, when properly installed, provide greater protection against elements of the environment than conventional reinforcement in a similar position. Poor workmanship, and more adverse environmental conditions that are not envisioned and accounted for at design and construction, can lead to exposure of the post-tensioning steel and anchorage parts to water/moisture. This can lead to the deterioration of post-tensioning components that are not intended to be water resistant.

Maintenance is commonly confined to periodical visual inspection of the concrete members that include post-tensioning. The focus of inspection is the integrity of the finished stressing recess and general signs of corrosion.

REFERENCES AND ACKNOWLEDGMENT

In preparing the material for this chapter, the following resources were reviewed/consulted.

- BBR post-tensioning system: www.BBRnetwork.com
- CCL post-tensioning: www.CCLint.com
- Freyssinet post-tensioning: www.Freyssinet.com
- Impacto: Fortaleza
- Oncrets: www.oncrests.com
- VSL post-tensioning systems: www.VSL.com

NOTE

1. In California the ratio is nearly 1; in Switzerland it can be as high as 7.

Index

ACI 318
   1-way slabs, 53, 33
   2-way slabs, 53, 93
   allowable deflection, 81
   allowable stresses, 83
   code requirements/design steps, 78
   column-supported floors, 53
   design flow charts, 84
   detailing, 89
   floor slab classification, 78
   load combinations, 82
   minimum rebar requirement, 112
   punching shear example, 123
Analysis
   finite elements, 16
   options, 54
   strip method, 15
   yield line, 196
Application
   beam and slab construction, 21
   column-supported floors, 19
   external post-tensioning, 25
   ground-supported slabs, 22
   high seismic risk regions, 26
   industrial floors, 23
   mat foundation, 22
   reaction modification, 28
   repair and retrofit, 25
   seismic deformation correction, 26
   slab on expansive soil (SOG), 22
   transfer floors, 24

Balanced loads
   load balancing, 35
Beams
   allowable stresses/stress check, 143
   beam and slab construction, 21
   code check for serviceability, 143
   crack control detailing, 143
   design moments, 148
   ductility check, 150
   effective flange width, 133
   example, 133
   one way shear design, 151
   strength design, 148
   strength requirements, 147
Bonded system
   application, 1
   construction, 219
   response to load, 1, 155
Building code
   ACI 318, 53
   ACI 318-EC2 differences, 56
   ACI-EC2 differences, 56
   ACI-specific provisions, 55
   code considerations, 53
   code objective, 55
   crack control, 59, 63
   deflection control, 54
   design steps, 53
   ductility, 55
   EC2 design steps, 58
   EC2-specific provisions, 55
   maximum reinforcement, 55
   minimum reinforcement, 55

Column-supported floors
   application, 19
   beam and slab construction, 21
   code compliance, 93
   design example, 93
   integrity steel, 128
   podium slabs, 20
   strip method, 14

242 Index

Comprehensive design method
  description, 48
Construction
  schedule, 230
Crack control
  allowable values, 55
  control through reinforcement, 195
Creep
  shortening, 181

Deflection
  allowable values, 108
  long-term, 109
Design
  ACI serviceability check, 108
  application of design methods, 50
  beam example, 93, 133
  capacity estimate, 114, 148
  code compliance, 53
  comprehensive method, 48
  cracking, 108
  design options, 36
  design steps, 53
  design strip, 93
  EC2 provisions, 35
  EC2 serviceability check, 105
  initial condition, 138
  integrity steel, 128
  load balancing, 41
  load combinations, 65, 77, 83, 88, 89
  load paths, 53
  minimum precompression, 55
  minimum reinforcement, 55
  moment capacity, 114
  one-way beams and slabs, 133
  punching shear, 119
  requirements, 35
  safety check, 114
  straight method, 40
  strength calculation, 116
  stress checks, 106, 110
  strip method, 14
  transfer of column moment, 128
Design strip
  detailed example, 93
  introduction, 1
Detailing
  banded-banded layout, 204
  banded-distributed layout, 199
  cantilever tendon profile, 205
  classification, 210

distributed-distributed
  layout, 202
effect of precompression, 196
flow chart, 212
live end, bonded, 221
live end, unbonded, 223
profile along wall, 210
profile at discontinuities, 208
profile at ramps, 208
profile at steps, 210
PT and RC difference, 195
tendon arrangement, 198
tendon detailing, 195
tendon profile, 205
trim bars, 210
Ductility
  requirement, 55

Earthquake design
  deformation restoration, 26
  flat slab construction, 26
EC2
  crack control, 97, 98
  cracking moment, 77
  design crack width, 63
  flow chart for serviceability, 68
  load combinations, 65
  punching shear example, 120
  requirements for post-tensioning, 53
  serviceability limit requirements, 67
Effective flange
  flanged beams, 133
Effective width, *see* effective flange
Elastic shortening
  stress loss, 180
Elongation
  computation, 178
  evaluation, 236
  measurement, 235
European Code
  European Code EC2, (2004), 53
Examples
  beam frame example, 133
  example for two-way system, 93
  podium floors, 93
  stress losses in prestressing, 185, 187

Finite elements
  application in building design, 1
Friction
  angular coefficient, 177

Index

coefficients, 177
draw in, 175
elongation, 178
examples, 183
formula, 177
friction loss diagrams, 175
seating losses, 175
wobble coefficient, 177

Ground supported slabs
  industrial floors, 23
  mat foundation, 22
  slab on expansive soil, 22
Grouted system, *see* bonded system
Grouting
  description and application, 237

History of post-tensioning
  application in building structures, 1
Hyperstatic actions
  calculation example, 147
  description, 35
  indirect method, 147
Hypothetical design stress
  ACI definition, 14
  EC2 definition, 60
  relationship between EC2 and ACI, 60

Industrial floors
  function and example, 23
Initial condition
  ACI example, 127
  EC2 example, 127
  load combination, 126
  requirements, 126
Integrity steel
  requirement, 128

Jacking 232; *see also* stressing

Load balancing
  concept introduction, 13
  example, friction, 183
  example, long-term loss, 185
  extended load balancing, 45
Losses
  long-term losses, 180
  relaxation in prestressing, 182
  simple load balancing, 44
  straight method, 40
  stress losses, 175

Mat foundation
  description and application, 22
Mesh
  mesh provision, 228
  mesh reinforcement, 228
Moment
  cracking moment, 58, 118

Non-prestressed steel
  maximum EC2, 66
  mesh reinforcement, 228
  minimum EC2, 66
  minimum two-way, 84

One-way systems
  ACI 318 requirements, 133

Podium slab
  definition, 20–21
Post-tensioning
  application in buildings, 20
  bonded system, 9
  comprehensive design method, 48
  concept, 3
  construction cost, 226
  construction cycle, 231
  cost, 224
  design method application, 50
  design method options, 40
  design requirements, 35
  development, 5
  distribution of precompression, 12
  hardware, 19, 219
  hyperstatic actions, 35
  introduction, 1
  load balancing method of design, 41
  precompression, 155
  quantities, 224
  response to member deformation, 11
  response to support restraint, 13
  shortening, 155
  simple design, 40
  steel, 35
  straight method of design, 40
  stressing operation, 232
  tendon arrangement, 227
  time of stressing, 232
  unbonded system, 5
  US history, 1
Precompression
  concentrated load support, 196

Index

diversion to wall supports, 169
impact on member strength, 196
impact of stiff walls, 171
member strength, 158
minimum required, 55, 55
in multi-story frames, 170
precompression/shortening, 155, 155
Prestressing
bonded system, 221
concept, 1
historical development, 1
live and dead end bonded, 221
live and dead end unbonded, 223
material, 35
prestressing steel, 219
prestressing strand, 219
quantities, 224
steel, 35, 219
system components, 219
unbonded system, 223
Pre-tensioning
concept, 3
Propping
requirement, 237
Punching shear
ACI example, 123
EC2 example, 120
requirement, 119

Raft foundations
application, 19
Rebar
details, 214
flow chart, 212
minimum required, 55, 68
trim bar plan, 213
Redistribution of reactions
description and example, 28
Reinforcement
detailing, 210
reinforcement, 55
trim bars, 210
Relaxation
relaxation in prestressing, 182
Repair and retrofit
description and example, 25
Restraint of supports
consequence, 155
cracks, 155
impact on moment capacity, 155

Safety
impact of support restraint, 177
requirement, 53
Secondary Moments, see Hyperstatic actions
Seismic design
application, 26
deformation control, 26
Serviceability check, SLS, see Service condition
Service condition, SLS
ACI requirements, 55
EC2 requirements, 55
Shear
one way shear, 151
punching shear, 120, 123
Shoring
shoring, 237
Shortening
creep shortening, 181
effect in multistory frames, 170
elastic shortening, 180
example, 169
outline and effects, 155
shortening and member strength, 158
shortening and precompression, 155
shrinkage shortening, 181
Shrinkage
cracking, 161
restraint of supports, 155
shortening, 181
Simple design procedure
straight method, 40
Simple frame method (SFM)
beam frames, 133
floor slabs, 93
Slab-on-Grade
function and example, 22
Slabs on expansive soil, see Slab-on-grade
Software
BIM-based, 17
finite elements, 17
introduction/application, 15
strip method, 15
SOG, see slab-on-grade
Strength evaluation
precompression effect, 158
section strength, 114
Stressing
ACI 318 allowable values, 127
EC2 allowable values, 127

elastic shortening, 180
elongation, 178
elongation measurement, 235
evaluation of elongation, 236
finishing stressing block, 238
friction loss, 177
hypothetical values, 14
losses, 175
loss to creep, 181
loss to relaxation of prestressing, 182
loss to shrinkage, 181
stressing equipment, 234
stress loss examples, 183
time of stressing, 232
Strip method
application in post-tensioning, 14
floor slabs, 93
software, 15
Structural system
beam and slab construction, 133
column-supported floors, 53
Support lines
background, 1

Tendon detailing
ACI 318 provisions, 78
anchorage detailing, 219
arrangement, 198
banded-banded layout, 204
banded-distributed layout, 199
bonded tendons, 9, 11, 221
cantilever profile, 206
common profiles, 205
different profiles, 205
discontinuities, 208
distributed-distributed layout, 202
end span profile, 99
installation sequence, 230
PT and RC detailing difference, 195
ramp modeling, 208
securing tendon profile, chairs, 227
tendon along walls, 210
tendon arrangement, 227
tendon at drop in elevation/steps, 209
tendon detailing, 195
tendon profile, 205
tests on arrangement, 198
unbonded tendons, 6, 223
Transfer floors
function and example, 24
Transfer of prestressing, *see* initial condition
Trim bars
required and details, 195
Two-way system
ACI 318, 53
floor example, 93

Unbonded system
hardware, 1, 219